AVIATION RECORDS IN THE JET AGE

The Planes and Technologies Behind the Breakthroughs

SR-71 RSO William A. Flanagan, Lt. Col., USAF (Ret.)

specialtypress
PUBLISHERS AND WHOLESALERS

Specialty Press
838 Lake Street South
Forest Lake, MN 55025
Phone: 651-277-1400 or 800-895-4585
Fax: 651-277-1203
www.specialtypress.com

Author Note: Some of the archival photos in this book are of noticeably lower quality. However, they have been included here because of their rarity, and historical importance in telling the story of record-breaking aircraft throughout history.

All photographs and artwork contained in this book are courtesy of the author, National Archives via Dennis R. Jenkins, and the Tony R. Landis and Mike Machat collections, unless otherwise noted.

Edit by Mike Machat
Layout by Monica Seiberlich

ISBN 978-1-58007-230-4
Item No. SP230

Library of Congress Cataloging-in-Publication Data

Names: Flanagan, William, author.
Title: Aviation records in the Jet Age : the planes and technologies behind the breakthroughs / William Flanagan.
Description: Forest Lake, MN : Specialty Press, [2017] | Includes index.
Identifiers: LCCN 2016040712 | ISBN 9781580072304
Subjects: LCSH: Airplanes–Technological innovations–History.
Classification: LCC TL671.2 .F55 2017 | DDC 629.133/349–dc23
LC record available at https://lccn.loc.gov/2016040712

Written, edited, and designed in the U.S.A.
Printed in China
10 9 8 7 6 5 4 3 2 1

Distributed in the UK and Europe by
Crécy Publishing Ltd
1a Ringway Trading Estate
Shadowmoss Road
Manchester M22 5LH England
Tel: 44 161 499 0024
Fax : 44 161 499 0298
www.crecy.co.uk
enquiries@crecy.co.uk

Front Cover: *North American's XB-70 Valkyrie, shown here in 1965 with drag chutes deployed, was the largest aircraft ever to fly Mach 3.*

Front Flap: *Douglas test pilot Gene May poses with the D-558-1 Skystreak on Rogers Dry Lake, California.*

Front Endpaper: *Convair's record-breaking Mach 2 wonder, the delta-wing B-58 Hustler, seen here landing on Runway 04 at Edwards AFB, California.*

Title Page: *Two Air Force test pilots paint their records on the Bell X-1A – Charles E. "Chuck" Yeager (on ladder) and Arthur "Kit" Murray.*

Table of Contents: *Lockheed's triplesonic SR-71 Blackbird set numerous world speed records, and is the highest-performance operational aircraft ever built.*

Rear Endpaper: *Holder of the World Absolute Speed Record in 1959, the Convair F-106 is the fastest single-engine jet aircraft ever flown, even to this day.*

Rear Flap: *Lt. Col. William A. Flanagan, USAF (Ret.) served as a flight test engineer on the Mach 3+ Lockheed SR-71 Blackbird, his favorite airplane.*

Back Cover Photos

Top: *Douglas's futuristic X-3 Stiletto looked like it was going Mach 2 just sitting on the ground. In actuality, it never reached that speed due to underperforming engines.*

Bottom: *The most successful research airplane ever flown, the rocket-powered North American X-15 set hypersonic speed records that still stand to this day.*

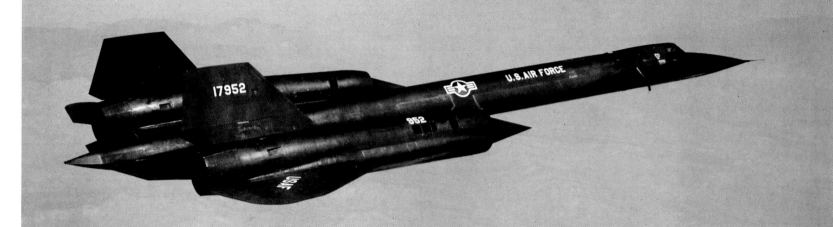

TABLE OF CONTENTS

PREFACE

In September 1924, three large biplanes landed at Clover Field in Santa Monica, California, having completed the first circumnavigation of the globe. This arduous flight took six months and was supported by hundreds of U.S. Navy crewmen manning supply ships all along the route. Those courageous airmen who flew these airplanes did not have the convenience of GPS: indeed, much of the journey was flown over water because no airports existed yet at most of the destinations along the route.

The Douglas World Cruisers represented the peak of aviation technology at the time. Only a short decade later, the first solo round-the-world flight occurred, taking less than one week of flying time in a single-engine monoplane. The next attempt in a high-speed twin-engine transport just a few years later took less than four days. In the late 1930s, myriad speed and altitude records were established, and international and interservice rivalries made these attempts even more fascinating to the general public, but how was all this possible?

This book tells the story of the highly advanced technology behind many famous record-breaking aircraft throughout history, focusing on the exciting years following World War II when new speed, altitude, and endurance records were set almost routinely. Famed aviation author Robert Serling once said, "Aviation progress was made one nut and bolt at a time," and in this book, Lt. Col. William A. Flanagan, a former crewman on the Mach 3–plus Lockheed SR-71 Blackbird, explains the compelling story of how manned flight was able to progress from a record speed of 300 mph to more than 3,000 mph in only 20 years' time.

As further evidence of this stunning rate of aviation progress, detailed appendices at the end of this book list major world speed, and altitude records, plus the aircraft and pilots that made them. From afterburners and rocket power to advanced aerodynamics, pressure suits, and digital design, this book gives you the inside story of famous record-breaking aircraft, and tells the fascinating tales of the men and machines that made aviation history.

Douglas Aircraft Company artist George Akimoto captures the excitement and anticipation of the crowd at Santa Monica's Clover Field as the Douglas World Cruisers prepare to depart for Seattle to begin the first transglobal flight in history, March 1924. Three of the five aircraft completed the trip in six months and 373 hours of flying time. (Wings & Airpower Historical Archive)

INTRODUCTION

At the beginning of the 20th Century, the first airplanes were flown by daredevils in search of prizes as well as wealthy sportsmen in search of thrills in 1909. John Moore-Brabazon was awarded British pilots license number one flying a Wright biplane, and felt sufficiently confident of his airplane and his flying skills to take an unusual passenger up for a ride.

The reference on the sign refers to a popular British and, later, American saying indicating disbelief by saying "when pigs fly." The photograph also indicates that early flying machines were literally fly-by-wire (as opposed to digital fly-by-wire today), as biplane wings had to be braced by numerous struts and wires. It seemed ludicrous at the time to believe that anything looking so fragile could change the world. But that is exactly what happened; aviation changed our world and literally expanded our horizons. In the United States alone, more than 30,000 flights occur each day carrying passengers and cargo. It took a long time to develop this network, starting with a 12-second 120-foot flight on 17 December 1903.

The path of this progression is marked by many milestones that are documented by aviation records. A look at the history of aviation records and the technology that made them possible gives us glimpses of the future of aviation in the 21st Century.

John Moore-Brabazon, 4 November 1909.

The Importance of Record Flights to Aviation Development

Records and the intense competition to achieve them benefited aviation in many ways: they fostered national pride, rewarded winners with prize money, and fed the natural human sporting drive to see who is best. In the early days of aviation, airplanes had such limited flight envelopes that they were treated as amazing oddities of little practical value, but thrilling to watch as amusements similar to horse races or polo matches. It is no accident that the organization established to maintain records, the Federation Aeronautique Internationale (FAI), evolved from a confederation of national auto clubs with members who were enthusiasts and sportsmen.

As airplanes improved and greater speeds and distances became possible, competition developed between designers and pilots, often in different countries. These competitions received widespread media attention, and the public flocked to see them and further understand the new aerial oddities. World War I occurred while airplanes were still relatively primitive, but by the end of that war aviation technology had progressed markedly and competitions were used to demonstrate the higher speeds produced by more powerful engines and the minimized drag of sleek new monoplanes that replaced World War I–era biplanes. During this era of rapid aviation progress, the fastest airplanes became matters of national pride.

Large cash prizes to the first aviators to cross oceans or to span continents further advanced the use of airplanes as a means of transportation, and the reliability of engines improved drastically to the point where the public not only watched airplane races, but actually used airplanes to go from point A to point B without fear of dying during the trip. World War II again stopped public record-setting, but by the end of that war new types of engines known as the turbojet and the rocket promised to allow much higher speeds than did the traditional piston-powered airplane.

In the 1950s, a Cold War between the capitalist West and the Communist Soviet Union furthered national competition; higher speeds and altitudes fostered national pride and demonstrated the potential military might of the opposing sides. Turbojet-powered commercial aircraft allowed more people to fly above the weather for greater distances than ever before in total comfort and with reasonable safety. The Cold War was fought with competition, rather than shooting, as exemplified by the race to the moon between the United States and the Soviet Union. Although expensive, very few lives were lost, and the technological advances inherent in the race completely revolutionized human life with the development of the advanced portable computers we take for granted today.

Computer technology expanded rapidly, as everyone knows who tries to keep pace with the latest version of their iPhone, and this may lead to aviation technology advances that seem as amazing to us today as the jet engine would have seemed to the Wright brothers. New aviation records will be set that may not be as headline-grabbing as breaking the sound barrier, but they will nevertheless affect the lives of humans far into the future.

FAI: The Keeper of Aviation Records

The world organization responsible for certifying aviation is the Federation Aeronautique Internationale (FAI) headquartered in Geneva, Switzerland. Established in 1905, it is a federated organization consisting of aeronautic associations in various countries throughout the world. The members of these aeronautic associations in turn act as agents of the FAI in certifying records established in their respective countries. One of the FAI's founding members was the Aero Club of France, established around the same time by members of the Automobile Club of France. In the United States the National Aeronautic Association is the national aero club and is the U.S. representative of the FAI.

Records are established for various classifications of aerial vehicles including balloons, land planes, seaplanes, helicopters, spacecraft, gliders, parachutes, ultralights, etc. Land planes is the largest category and is further subdivided by type of propulsion (piston engine, turboprop, jet engine, and rocket engine). Land plane categories are also further subdivided by weight so that records may be established depending on the size of the airplane. This is so a Piper Cub does not have to compete with a Boeing 747 for the same record. To set a new record, the old record must be exceeded generally by 1 percent for speed and 3 percent for altitude.

A special category of records is the Absolute World Records for heavier-than-air vehicles. These include Altitude, Speed Over A Straight Course, Speed Over A Closed-Circuit Course, Altitude In Horizontal Flight, Great Circle Distance Without Landing, and Distance In A Closed-Circuit Without Landing. Record-setting was officially suspended by the FAI during World War I and World War II for secrecy reasons.

All of the records that are international in scope, and necessarily the best world performance for a particular type of aviation, are termed World-Class Records to differentiate from the Absolute World Records. In addition, the national aero clubs generally track records within their respective countries, thereby recognizing each country's aviation and space records.

Although the records that attract the most attention are the Absolute Records for speed and altitude, in today's world the most popular records to be set are those for "speed over a recognized course" with the "recognized course" usually being various cities throughout the world. Aerial commerce now spans the globe and many city pairs have airline service with ground speeds occasionally approaching 600 mph. This allows new records to be established as a by-product of normal aviation commerce.

The National Aeronautic Association (NAA) of the United States' website has very careful instructions on what is required for a pilot to set a record. It requires preplanning, filling out forms with the record intended to be set, the proposed dates on which the record will be set, necessary documentation, and of course a fee to cover expenses for the documentation. The NAA will then issue a 90-day license for the proposed record to be set. Within 3 days of actually accomplishing the event the NAA must be notified along with proper documentation to prove that the event occurred. The NAA will then notify the FAI, and after due verification, certificates are mailed to the new record holder.

Record-setting is considered a sporting event by the NAA and FAI, and an annual awards banquet is generally held in Washington, DC, where the certificates can be presented, if desired.

Chapter 1

AVIATION RECORDS THROUGH WORLD WAR I

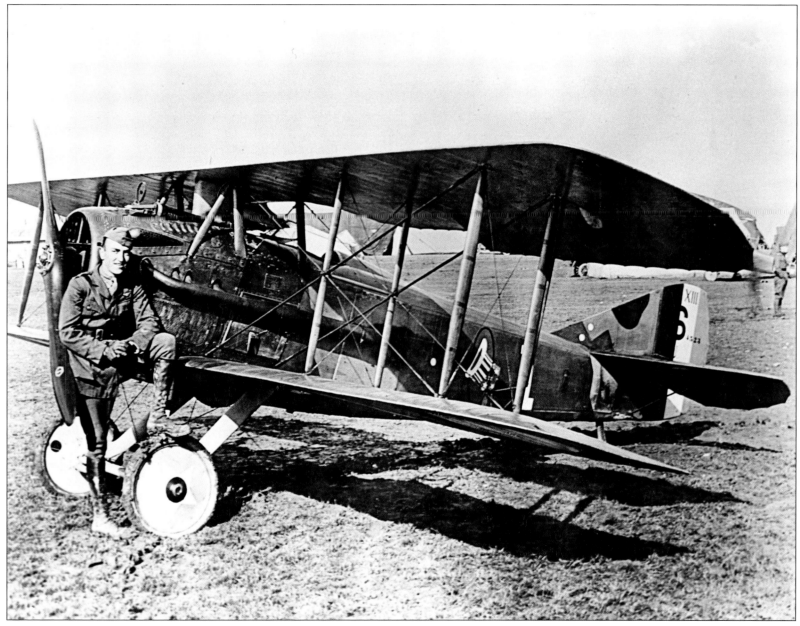

Considered the "Stealth Fighter" of its era, the French SPAD represented the ultimate advancement in high-performance flight in 1918, a mere 15 years after the Wright brothers' first flights. Shown here with American Ace Capt. Eddie Rickenbacker, a former racing driver turned pilot, the SPAD biplane fighter set an impressive speed record of 191.98 mph on 4 November 1920 flown by French pilot Baron de Romanet. (Wings & Airpower Historical Archive)

All objects on Earth are subject to a common natural law: Gravity pulls everything toward the center of the planet with a force that results in an acceleration of 32 feet per second in freefall. In addition, Earth's gaseous atmosphere is composed mainly of nitrogen and oxygen that, at sea level, exerts a pressure of approximately 15 pounds per square inch. These two facts are both a blessing and a curse for the destiny of our planet in general, and are the key elements to the development of what we call "aeronautics."

Development of the Heavier-than-Air Flying Machine

The Wright brothers' story is well known but what is not always realized is that Orville and Wilbur were businessmen from Dayton, Ohio, who as bicycle shop owners were interested in the development of an airplane for practical transportation as well as the accomplishment of simply flying. As Wilbur said in a speech in 1908, "When Orville and I began, we had no idea whether flying was going to be possible or not, but what we had made up our minds to do was to find out." By 1902 when they had perfected gliders that could actually turn they began to realize the value of

Wright glider of 1902 flying at Kitty Hawk. This was the first design to allow wing warping combined with the rudder to successfully turn the aircraft. The Wright brothers practiced flying using this vehicle and based upon their success, they determined that when they returned in 1903 they would have an engine installed on the glider and attempt powered flight. (Public domain via NASM)

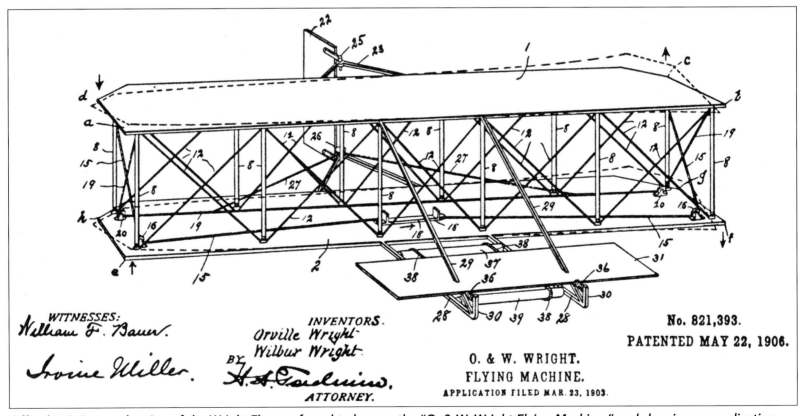

Official U.S. Patent drawing of the Wright Flyer, referred to here as the "O. & W. Wright Flying Machine," and showing an application date of March 23, 1903. Critical to the success of the Flyer was its ingenious control system, using mechanical cables and pulleys to warp the outer wing surface, thereby changing the airfoil camber and effecting a bank angle with which to turn the aircraft. Patent was granted on 22 May 1906. (Wings & Airpower Historical Archive)

their airplane as a serious business proposition. For that reason they were more guarded about their progress as they started the arduous process of applying for a U.S. patent. Their flight test base was on the Outer Banks of North Carolina due to the need for prevailing strong breezes to help their gliders get airborne.

By installing a 4-cylinder gasoline engine developing a meager 12 hp and using it to drive two propellers via bicycle chain drives, on 17 December 1903, their heavier-than-air machine took to the sky under its own power for the first time, successfully landing at the same height from which it started. A terse telegram (Western Union charged by the word) to Bishop Wright announced the morning's accomplishments and ended with the cryptic words "inform press home Christmas Orville Wright." The press was hardly impressed and few newspapers announced it. The British *Daily Mail* devoted 2 inches of text on 18 December calling the flyer a "balloon-less airship," which somewhat concealed the significance of the event.

The Wright brothers adjourned to Dayton by Christmas and did not fly in public for another two years as they studied the results of their initial flights and experimented with new designs at Huffman Prairie in Dayton. Thus, the accomplishments of 17 December remained largely a secret except to close friends of the Wright brothers.

The Wright Brothers Take Europe by Storm

Meanwhile in Europe the center of aeronautical development remained in Paris, France, where it had been since the Montgolfier brothers first flew balloons in the 1780s. The intrepid Brazilian aviation sportsman Santos Dumont now turned his attention from dirigibles to flying machines. His initial airplane was known as the 14bis in honor of the fact that it was originally launched from a dirigible of his entitled 14, albeit as a glider. His flights over public parks in Paris in late 1906 were recognized by the newly established FAI (see page

Basic Aerodynamics: Lift and Drag

Four forces work on a body in flight: weight, lift, drag, and thrust. Weight is a characteristic of the body; lift is the force that must overcome the weight in order to fly. Drag is the force the atmosphere exerts on the body to resist its motion; thrust is the force the body must generate to keep moving through the atmosphere.

Lift is generated by a surface attached to an aircraft's body called a wing or airfoil. The wing is shaped in such a way that as the vehicle moves through the air, atmospheric air pressure on top of the wing is less than the air pressure on the bottom of the wing, producing a net force that allows the wing to rise and the vehicle to leave the ground.

The coefficient of lift is determined by the angle of attack, or the angle between the wing and the flight path of the airplane. The higher the angle of attack (higher nose-up attitude), the higher the coefficient of lift up to a certain point known as the stall. Beyond the stall, the lift coefficient decreases, causing the aircraft to descend now that the force of lift is lower than the aircraft's weight.

The faster an airplane goes, the more lift it generates. Unfortunately, the faster an airplane goes through the atmosphere, the more drag force it must overcome to maintain the speed. A similar equation to lift can be written to represent the drag force on the airplane. The coefficient of drag values are usually smaller than the coefficient of lift at low speeds, but there are different types of drag. Profile drag is that caused by the frontal area of the body, which is moving through the atmosphere; induced drag is caused by the lift force on a wing; and skin friction, or parasitic drag, is caused by air moving over the body from nose to tail. At high speeds near or above the speed of sound there is an additional type of drag called wave drag or compressible flow drag due to shockwave formation.

In order to overcome drag and remain airborne at a given speed, thrust is necessary to maintain velocity. The thrust in turn requires a power plant, and the development of relatively lightweight gasoline engines toward the end of the 19th Century meant that the era of human flight was possibly imminent. Initially experimenters concentrated on unpowered gliders to develop designs that could fly only short hops, but at least they could leave the ground. The noteworthy experiences of glider pioneers such as Otto Lilienthal in Germany and Octave Chanute in the United States received publicity and much attention from aviation enthusiasts. The Wright brothers in the United States corresponded with Chanute and eventually embarked on a campaign to add a power plant to a glider, thus developing a powered, controlled airplane.

Wright Glider demonstrates the properties of lift and drag.
(Mike Machat)

6) as the first flights by a powered airplane in Europe. Santos Dumont with his first flight won the Archdeacon prize of 3,000 francs for the first airplane in Europe to fly more than 25 m (about 80 feet). For the second flight, he won the 1,500-franc prize offered by the Aero Club of France to the first plane to fly more than 100 m (330 feet).

This public performance should be contrasted with the fact that the Wright brothers back in Dayton were now routinely flying for up to 30 minutes, observed only by passengers on the local trolley line. Nevertheless, this public performance rekindled interests in aviation in Europe and, later, the world. The financial rewards offered by the Aero Club of France doubtless also sparked many sportsmen to raise the bar for future flying. The publicity also spurred the Wright brothers to make the decision to begin demonstration flights before the world in 1908. The awarding of a U.S. patent in 1906 also helped raise the curtain of secrecy. The Wright brothers were about to step onto the world stage.

The Airplane Arrives on the International Scene; Records Are Set

The elements of airplane design now began to take shape simultaneously in the United States and Europe. In Europe, the honors were split between France and England, with the most well-known names being French. Another pair of brothers, named Voisin (Gabriel and Henri), started building gliders before 1903 and specialized in biplanes and triplanes. They became the first commercial manufacturers of airplanes in Europe, providing airplanes primarily to sportsmen. Using their airplanes, Henri Farman, a champion bicycle rider and race car driver, set a record in 1907 when he flew 770 m (2,530 feet) in 52 seconds and at a speed greater than 50 mph, thus establishing himself as the fastest flyer in the world. In October 1908, Farman made the first cross-country flight in Europe by flying 17 miles to Reims in 20 minutes. But the aviation picture was about to change: America's Wilbur Wright, and later the Wright's arch rival, Glenn Curtiss, were about to arrive.

The Wright Model A was similar to the 1905 Flyer 3 developed in Dayton, but with several significant differences, such as upright seating for the pilot and accommodations for a passenger. Because the Wright brothers were now engaged in selling their airplane, Orville conducted demonstrations for the army at Fort Meyers in Washington, DC, while Wilbur took a Model A to France for demonstrations there. No particular attempts were made to set records for the military in the United States, but Europe needed a demonstration of the performance of their machine because many people there still didn't believe the Wright brothers had actually flown.

The fateful demonstration day occurred on 8 August 1908 at the racetrack near Le Mans. Wilbur took off in front of a crowd of aviation enthusiasts, made five perfectly controlled circles, and then landed right in front of the crowd. He subsequently gave multiple demonstration flights, including those carrying passengers. On 31 December, he set an endurance record of 2 hours and 20 minutes in

the air as well as setting an altitude record on 18 December of 110 m, or 360 feet. These records were all officially certified by the FAI and thus, in one historic series of flights in 1908, Wilbur Wright almost doubled distance records, almost quadrupled the endurance record, and quadrupled the altitude record.

As a result of this tour the European aviation industry morphed from Wright detractors to the Wright brothers' fan club. The records were there for all to see; as one French aviator said at the time, "it is true; we are beaten." Construction licenses for the Model A were sold to Great Britain, France, and Germany; the prizes and trophies may have gone to the Europeans, but the Wright brothers cared more about contracts than honors. As icing on the cake, in the United States, Orville's demonstration had also been successful in convincing the U.S. Army to buy airplanes. Although the demonstration was marred by a fatal crash that killed army observer LT Thomas Selfridge, the Model A flew well.

The airplanes built for the army contract actually flew in 1909, and the Wrights were paid $25,000 for the plane plus a $5,000 bonus for a 2½-mph speed increase over the army's 40-mph specification. It could be said that 1908 was the year in which the world first realized that the age-old dream of flying in a machine through the air had been achieved; 1909–1911 were the years in which aviation mania appeared. Records were established and quickly broken in these years as aviation matured at an incredible pace. And now one of the Wright brothers' biggest competitors was to become another American, by the name of Glenn Curtiss.

The Wright Brothers and Glenn Curtiss

The Wright brothers' success at Kitty Hawk on the Outer Banks of North Carolina, attracted the attention of many intellectuals, one of whom was the famed American inventor Alexander Graham Bell, who formed an organization known as the Aerial Experiment Association (AEA) with the objective of conducting further aerial research. While forming the association he met a prominent bicycle racer and motorcycle designer, Glenn Curtiss, who suggested he could assist Bell in his endeavor. The AEA produced initially four aircraft, one of which, the *June Bug* biplane, was essentially designed by Curtiss. As a member of the prestigious AEA he had been invited to look at the Wright brothers' 1905 design and he immediately recognized the significance of the banking-to-turn design of the wings of the Wrights' biplane.

During their studies of the problem of designing an airplane the Wright brothers had recognized that by making the vehicle too stable it would be almost impossible to turn. It was therefore deliberately designed to be inherently somewhat unstable like a bicycle, requiring pilot attention to keep it on a chosen path but making it easier to turn. The wing warping of the 1903 design allowed the pilot to keep it level as well as steer by shifting his weight in a cradle as he lay prone on the lower wing.

Sister Katherine Wright carefully observes army personnel preparing to hangar the "B" Flyer during its flight demonstration tour in Europe. By December 1908, distance flights of up to 80 miles and endurance flights of more than two hours had been successfully accomplished, a stunning achievement for only five years after the first flights of the original Wright Flyer. (Wings & Airpower Historical Archive)

As a bicycle and motorcycle racer himself, Curtiss understood the necessity of constant pilot attention to controlling a practical machine. The Wright patent granted in 1906 outlined the use of wing-warping to allow the airplane to turn. To avoid infraction of the patent, the Curtiss design used individual panels on each wing rather than bending the entire wing. This unfortunately led to a long-standing dispute between the Wright brothers and Curtiss on whether or not the 1906 patent had been infringed upon. Using ailerons, on 4 July 1908 the *June Bug* won the first aviation prize to be awarded in America (the Scientific American trophy) by flying nearly 1 mile in less than 2 minutes before official observers of the American Aero Club.

When the AEA dissolved, Glenn Curtiss began designing and selling very successful biplanes that soon were matching the performance of Wright airplanes. The 40-hp V-8 air-cooled engine of *June Bug*, an adaptation of a Curtiss motorcycle engine, proved to be an effective and highly reliable engine. Its reliability was proven in the new Curtiss biplane *Gold Bug* that captured the second Scientific

American contest by flying more than 25 km on 17 July 1909. The same machine was used for a long-distance flight on 31 May 1910 from Albany to New York City, winning for Glenn Curtiss the New York World prize of $10,000. The flight covered 142.5 miles and was completed in 2 hours and 50 minutes. This same flight also earned Curtiss his third Scientific American trophy.

In Europe, attention was shifting to monoplanes with only one wing as opposed to the American biplanes. The intrepid Santos Dumont in 1909 developed a tiny monoplane called the *Demoiselle*. The airplane was actually made for sporting pleasure and was Dumont's favorite airplane. It was so small that he transported the *Demoiselle* to his takeoff location in the back of what today would be called a pickup truck, complete with wings and engine. A further monoplane design was the *Antoinette*, a delightfully graceful design that looked like a bird, even while sitting on the ground. This could carry two people and was obviously built for sturdier missions than was the *Demoiselle*.

In this amazing photo, you see the preparation for takeoff of the Wright "B" Flyer at Centocelle Field southeast of Rome, Italy, on 15 April 1909. This photograph, the first ever created showing an airplane on the ground from above, was taken by a photographer aloft in a hot-air balloon. Visible at right is the launching catapult tower; the aircraft is sitting atop its launching rail assembly. (Wings & Airpower Historical Archive)

U.S. Army specifications stipulated acceptable methods for ground handling and mobility, including transportation of the aircraft by wagon. This requirement called for such structural components on the Flyer as landing skids and skid braces that could be folded outboard and stowed as shown here. (Wings & Airpower Historical Archive)

Pilot Harry Atwood takes off from the South lawn of the White House on 14 July 1911 in a Wright Model "B" after being presented the National Aero Club's gold medal by President Howard Taft. Atwood had flown from Boston to Washington, DC, the longest flight ever made in the United States at that time. Note the large number of interested spectators on the White House lawn.

With its 60-hp air-cooled engine it set several records in 1909 for speed and altitude flown by Paris-born Englishman Hubert Latham and was chosen by him to make the first English Channel crossing. Unfortunately, the reliability of the *Antoinette's* engine proved to be its weak link and twice resulted in him landing in the English Channel, the second time only one mile short of the White Cliffs of Dover. Instead, the honors and $5,000 *Daily Mail* prize went to Louis Bleriot, a French monoplane designer and pilot who was initially known more for his frequent crashes than his success at flying.

Bad weather had prevented both Latham and Bleriot from taking off, but a sudden break in the weather in the early morning of 25 July 1909, allowed Bleriot to take off while his rival Latham slept. Bleriot's wife watched the takeoff from France and then boarded a French navy destroyer, which followed him to the White Cliffs of Dover, arriving almost as soon as the airplane. The entire continent was agog over this aerial crossing of the English Channel, but especially in France, where Bleriot was idolized to an extent not to be seen again until the "Lone Eagle" Lindbergh arrived in Paris in 1927. At the Reims (France) International Air Exhibition in August 1909 Bleriot raised the official speed record to 48 mph. He became a world-renowned aircraft manufacturer and his monoplane designs were used not only for sport but were the first airplanes to be purchased by the French army and later by the British.

The excitement of 1909 with the crossing of the English Channel and the new speed records continued in 1910, which could be known as the Era of the Air Exposition. The Reims air show in

Official Program Friday, Jan. 14 1910

FIRST IN AMERICA
AVIATION MEET
LOS ANGELES
JANUARY 10-20 1910
American & Foreign Aviators
DAILY FLIGHTS
SOUTHERN CALIFORNIA DAY
PRICE 10 CENTS

August of 1909 had featured a 20-km race with a cup donated by Gordon Bennet to the winner as well as a cash prize. The cup was won by Glenn Curtiss at a speed of 47 mph with Louis Bleriot being behind by only 0.1 mph. The aerial exposition attention shifted in January of 1910 to Los Angeles. As the first American air exposition, it was far more successful than originally planned with more than 337,000 spectators present for the 10-day air show.

Aviators from France and England had their airplanes shipped to California to compete for the large prize money for speed and endurance that was awarded at the air show. Designs from the Wright Company as well as the Curtiss Company were popular as was a Farman biplane flown by Frenchman Louis Paulhan. On the third day of the meet he set a new official altitude record of 4,000 feet. The air meet was held near Los Angeles in an area called Dominguez Hills, near where Los Angeles International Airport is located. The success of the air meet fired enthusiasm for the airplane in America and actually was a trigger for Southern California to become extremely air minded.

The year 1910 ended with the Belmont Park aerial tournament in the first week of October in New York, described as "the top sporting event of the decade." Once again aviators from all over the world came to the United States to participate in, among other things, a race around the Statue of Liberty with a prize of $10,000 as well as competitions to fly higher than other competitors. By the end of the meet the world altitude record had been raised to 10,000 feet and the speed record exceeded 100 kmph (62 mph) for the first time at 109 kmph (68 mph). Perhaps equally significant to illustrate the

Official poster for the celebrated 1910 aerial exposition at Dominguez Hills south of Los Angeles, California. This was the first aerial meet held in the United States and was rousingly successful judging by the crowds in the grandstand on a weekday. Many famous European flyers attended, often flying Wright and Curtiss airplanes. The success of this air meet led to further shows in other American cities. This, in turn, led to the public finally accepting the fact that man indeed was now flying airplanes, and as in Europe, Americans became noticeably more aviation minded. (The Museum of Flying)

Lineup of Sopwith "One and One-Half Strutter" pursuit planes on the Western front. Initially developed as a fighter, with long-range bomber and escort aircraft variants, this rugged biplane was the first British aircraft fitted with new synchronizing gear that allowed the machine guns to fire directly through the spinning propeller blades. Deliveries began to the Royal Naval Air Service in early spring 1916. Military aircraft were now mass-produced and no longer fragile playthings. (The Museum of Flying)

acceptance of aviation was that in October, former President Theodore Roosevelt became the first resident of the White House to fly in an airplane at the St. Louis air meet. And in the following year a Wright biplane landed on the South Lawn of the White House for the pilot to accept an award from President Taft. Aviation was becoming serious business.

In addition to speed and altitude records, which were easily understood by the public, endurance and distance performance was also increasing rapidly. Endurance records were usually established by flying close to the takeoff point and staying aloft for as long as possible. Distance records, however, had to be established by traveling from a point to another point, and landing at the new destination or possibly returning to the original takeoff for what became known as a closed-circuit record. Record categories were created for minimum time from city to city; this was considered a better test for the sportsman to demonstrate his airmanship as he battled with changing winds and flying over unknown terrain.

Newspapers and philanthropists often awarded prizes for flights between destinations, with the winner being the aviator who achieved the flight with the minimum time en route. In this way, endurance records were set by flying slowly with best fuel economy and staying aloft as long as fuel was available. The race between destinations, on the other hand, required flying at a higher speed, then landing, quickly refueling, and getting airborne again as soon as possible. An example of a distance record was Harry Atwood flying from St. Louis to New York (1,266 miles) in 28 hours and 53 minutes flying time.

A truly monumental long-distance flight began from New York on 30 September 1911. The pilot was Cal Galbraith Rogers; the airplane, a Wright model EX, itself a modified Wright model B with a 35-hp engine. The objective was to fly the *Vin Fiz* (named for a popular soft drink) from the East Coast to the West Coast in less than 30 days, and thus win a prize of $50,000 offered by publisher William Randolph Hearst. Rogers followed the railroad tracks on which his support train loaded with spare parts was traveling. Because of adverse winds, sometimes the train traveled faster than the airplane. The route was 4,300 miles long, due to having to avoid obstacles such as the Rocky Mountains, over which the airplane could not fly.

Rogers departed Sheepshead Bay in Long Island on 17 September 1913 and arrived in Pasadena on 3 November. He had flown for 82 hours with an average speed of 52 mph and made 70 landings en route. Many of the landings were better described as controlled crashes; Rogers had only 60 hours flying time total when he departed on his epic journey. As a result, he spent time in the hospital recovering from injuries suffered during some of the landings. Very little remained of the original airplane in Pasadena because so many parts and even engines had been replaced during the journey. The airplane is now in the National Air and Space Museum in Washington, DC. Rogers won much acclaim for his persistence if not his landing skills, but unfortunately died in an airplane crash only four months after his cross-country trip.

The Great War Looms and Military Aviation Takes Center Stage

At the beginning of World War I airplanes were still somewhat in the category of curious toys, but that rapidly changed. The Italian air force had led the way during the Balkan Wars of 1912–1913 where airplanes were used both as aerial scouts to discover the intentions of armies on the ground and as menaces harassing enemy military formations with handheld bombs dropped from the cockpit of the reconnaissance airplane. In the Great War by 1915, 400 airplanes were engaged in military activities. Their importance magnified as trench lines were developed and settled from the North Sea to Switzerland. The importance of reconnaissance soon led to the development of a new type of airplane known originally as the scout and later as a pursuit or fighter plane designed to destroy the opposing force reconnaissance airplanes and deny the skies over the battlefront to the enemy.

Initially, rifles and pistols were used to shoot at enemy airplanes, with limited success, leading to the mounting of rapid-fire machine guns on airplanes. This installation presented its own problems as there was often a whirling propeller at the front of the airplane. Solutions included redesigning airplanes with pusher propellers mounted behind the pilot as well as having a second crew member to man the machine gun. Later, mechanical synchronizers were devised to shut off the flow of bullets as the propeller passed in front of the gun muzzle. The battle for air superiority over the trenches became the popular image of the use of the airplane, so much so that during World War I and even today, the pilot ace is the predominant, and often only, flyer portrayed in the Great War.

This view of World War I is a vast oversimplification of the influence of the war on aviation technology. By the end of the war there were almost 13,000 fighting planes in service. So important had become the struggle for air superiority that new fighter plane designs were coming out every year that made the previous year's airplanes death traps for the pilots that flew them against the newer airplanes. The number of airplanes built between 1914 and 1918 exceeded 175,000 versus the less than 10,000 airplanes that had been built in the world between 1903 and 1914. Aircraft building became a major industry throughout the world, and like the automobile, mass production of airplanes led to greater efficiencies as well as lower prices.

America did not enter World War I until 1917, and the Allies decreed that America's production capacity would be most useful to the war effort, resulting in America being assigned to build a Curtiss trainer airplane known as the JN-4D (Jenny) and a reliable 400-hp engine known as the Liberty, while American army air service pilots flew French and British fighter planes in combat.

What is lesser known today than the change in fighter planes was the change in bomber airplanes. Whereas early use of bombs was restricted to small handheld bombs, by the end of the war twin-engine and in some cases four-engine airplanes were designed to carry bomb loads of over a ton long distances to strike enemy targets, including cities and ports. These developments in fighter and bomber airplanes meant that aviation was on the threshold of an era where literally the sky was the limit.

With the arrival of peace in 1920, the FAI announced that they would open the competition for world records again in aviation. The same forces that drove the competition for records prior to World War I were in full play again. But now they were restarting from a higher level, thanks to the developments made during the war, especially in the development of higher powered engines. But attention to records now included not just top speed, but also altitude and distance. Now categories had to be established for the type of airplane, as well as the number of engines. And whereas previous records had been regarded as primarily sporting events, aviation was about to enter an age in which sporting performance took second place to practical use.

Mammoth Martin MB-2 biplane bomber flying over our nation's capital, circa early 1920s. Note the Washington Monument in the background. The absence of military flying apparel suggests a civilian factory flight test crew posing for a publicity photo; the Martin plant was only a short flight away near Baltimore. MB-2 had a crew of four, a top speed of 99 mph, and could carry a bomb load of 2,000 pounds over a range of 500 miles. The aircraft was also designated as the NBS-1 for "Night Bomber/Short Range." (San Diego Air & Space Museum Historical Archives, used with permission)

Chapter 2

THE GOLDEN AGE OF PROGRESS (1919–1939)

Taking off from Burbank Airport north of Los Angeles, Howard Hughes pilots his specially modified Lockheed 14 Super Electra as it undergoes extensive testing in preparation for his 1938 round-the-world record attempt. Flown eastbound from New York's Floyd Bennett Field (adjacent to what is now JFK Airport), the flight successfully circumnavigated the globe in just 3 days, 19 hours, and 17 minutes, shattering the previous record of 7 days and 19 hours set by Wiley Post in a modified single-engine Lockheed Vega in 1933 (see page 24). (Wings & Airpower Historical Archive)

This rare and historic, but poor-quality, photo shows the Deperdussin Monocoque being towed on the water at Monaco where it won the 1913 Schneider Cup. This diminutive floatplane racer was powered by a robust 160-hp engine, and was considered a large aircraft for its era, sporting a 44-foot wingspan. Note the rather large tail float, a common feature on early seaplane racers. (NASM)

Seaplanes are comprised of two types: flying boats and floatplanes. The flying boat's fuselage is actually shaped like a hull; the floatplane is a landplane with twin pontoon floats instead of wheels. Flying boats placed high in the early Schneider Cup races, but as speeds increased with more powerful engines, they soon lost out due to the increased drag at higher speeds of their hull-shaped fuselage. The Sopwith Tabloid (shown) is a floatplane, illustrating the smaller fuselage designed for flight rather than navigating on the water. (Wings & Airpower Historical Archive)

Judging by today's comparatively glacial evolution of any new military aircraft program, it's nearly incomprehensible to think that only 10 years after the Wright brothers first flew in 1903, manned biplane fighter aircraft were preparing for aerial duels in the darkening skies over Europe. Ten short years after that, aircraft had circumnavigated the globe, an astounding achievement in its own right. Now something called "air racing" dominated the headlines and new world records were broken almost as fast as they could be set.

The Racers: Speed Records Resume After the War

Beginning in 1912, the French developed a new type of airplane intended not only as a racer for air meets, but to achieve recognized speed records as well. These new airplanes used a construction method called monocoque design. A Swedish engineer named Ruchonnet proposed that, theoretically, a hollow fuselage shell made of plywood might be lighter, stronger, and more aerodynamically efficient than a conventional braced-girder fuselage structure. The Deperdussin firm decided to build a monoplane racer using such a plywood shell, and at the Reims air meet, it raised the world speed record to 203 kmph (126 mph), the first speed record to exceed 200 kmph (124 mph). World War I had ended world speed records for the time being, but the monocoque fuselage was now the wave of the future.

On 6 January 1920, the FAI announced they would begin accepting applications for records immediately, but that the rules had changed from prewar times. The speed records would represent the *average* speed achieved in four runs over a 1-km course to remove

wind effects. The first record under these new rules was set on 7 February 1920 at Villacoublay at 275 kmph (171 mph). The French pilot, Joseph Lecointe, flew a radical modification of the Newport 28 fighter. Using this airplane in 1920 raised the official speed record to more than 300 kmph (186 mph) for the first time, reaching 313 kmph (195 mph) on December 20. For this achievement, Lecointe received the gold medal of the Aero Club of France.

Races, Race Courses and the Schneider Cup

In America, postwar races were dominated by the Pulitzer competition. From 1922 to 1925 this event was held as a four or five lap race over a 30-mile course, sponsored by the Aero Club of America. Originally intended for military pilots, the race offered cash prize awards in the form of Liberty Bonds plus plaques donated by Ralph Pulitzer. Initially the airplanes were to be standard fighters developed for the war and now flown by the U.S. military. One exception was the Verville VCP-R purpose-designed racer built by the U.S. Army as an experimental vehicle and also intended to demonstrate American high-speed technology. The traditional army-navy competition, which started in aviation with the Pulitzer races, doubtless led to financial support within the services that would not have been forthcoming without the traditional "beat army/beat navy" goad.

America's Curtiss Company dominated the races in the development of engines and streamlined biplanes. In 1922 the army-navy competition resumed with a vengeance as the navy entered two improved Curtiss CR-2 racers, while the army entered two improved Curtiss R6 racers. The army airplanes came in first and second and

The army's sleek Curtiss R2C with its 507-hp engine is representative of the type of airplanes that dominated the Pulitzer Cup races. The navy finished first and second flying R2Cs in the 1923 Pulitzer race and a month later, new absolute speed records of 417 and 429 mph were set by the same two navy pilots who had won the Pulitzer race. Army LT Cyrus "Cy" Bettis flew an improved version of this landplane racer to a first-place finish in the 1925 Pulitzer Race.

The first practical demonstration of global airpower projection was made by the U.S. Army in 1924, using five Douglas biplanes to circumnavigate the Earth. Named the Douglas World Cruisers, these large, rugged aircraft not only made the first round-the-world flight in history, but also the first transpacific flight and first westbound transatlantic crossing. Powered by 420-hp Liberty engines, the World Cruisers departed from Seattle in March 1924, flying 28,945 miles in 371 flight hours over an elapsed time of 175 days, or nearly six months of continuous travel. The average flying speed was 78 mph. (Wings & Airpower Historical Archive)

U.S. Army LT James H. Doolittle, already a rising star in aviation circles, stands on the right float of his Curtiss R3C-2 seaplane racer at Port Washington, New York. "Jimmy" Doolittle won the 1925 Schneider Cup race in this aircraft, and then went on to a long and distinguished military career. He is perhaps best known for making the first "blind" flight of an airplane using only instruments, and leading the heroic Doolittle Raid on Japan in April 1942, the first U.S. attack on the Japanese mainland. That mission proved to be a turning point in the Pacific Battle of World War II. (NASM)

The Italians and British exchanged victories in the Schneider Cup races with increasingly streamlined airplanes possessing engine horsepower that grew from 800 to 1,900 in just three years. The Supermarine S4 (shown) is predecessor to the even faster S5 and S6 designs that won twice, but the Italians were hot on their heels and planned a super racer, the Macchi MC 72, for 1931 with twin 12-cylinder engines mounted in tandem producing 2,800-hp and driving contra-rotating propellers. It could not be completed in time to enter the race, so the British S6B flew uncontested with a 2,300-hp engine capturing the trophy. (NASM)

Streamlining and the Quest for Speed and Range

Drag force on an airplane moving through the air increases with the drag area and with the square of velocity (an airplane traveling twice as fast has four times the drag). In 1915, creation of the American National Advisory Committee for Aeronautics (NACA, forerunner of NASA) triggered concentrated studies to not only improve the lift capability of wings, but also to reduce the total drag of an aircraft. One of NACA's greatest contributions was the development of the NACA cowling, intended to reduce the drag of new multi-cylinder aircraft engines being developed while at the same time making them more efficient.

When moving through the air, a flat plate produces the most drag but a cylindrical shape is not far behind due to the turbulent disturbed flow on the downwind side of the cylinder. A teardrop shape, somewhat like a wing, smooths the disturbed airflow and reduces the drag force at higher speeds. NACA studies showed that by placing a close-fitting cowling over the radial engine with simple cowl flaps aft of the engine, drag could be significantly reduced. This discovery was rewarded in 1929 with the Collier Trophy, which today is still awarded annually to a person, company, or engineering team in the United States who has made a significant contribution to aviation in that year.

Another significant contributor to drag was landing gear. The cylindrical struts and wheels (although sometimes hidden by streamlined fairings, or "wheel spats") were a significant source of drag, especially at speeds above 200 mph. The Dayton Wright racer of 1920 was the first airplane with retractable landing gear. Once airborne, the pilot manipulated a hand crank to retract the wheels. This was adequate for a racer, but as airplanes grew larger and the weight of the landing gear became heavier, it was necessary to employ hydraulic systems (already installed for braking action after landing) to retract the landing gear.

In 1935, the single-engine H1 racer designed and built for Howard Hughes set a new world speed record for land planes of 567 kmph (352 mph) using a 14-cylinder radial engine that produced 1,000 hp, versus the 2,000-plus hp of the Schneider Cup speed racers. The H1's high speed with relatively low horsepower was an indication of Hughes' success producing a low drag airplane. He designed the fairings to reduce turbulent flow where the wings met the cylindrical fuselage, further reducing drag at high speeds. For setting absolute speed records, Hughes had a set of lower-aspect-ratio wings, but using the same aircraft, he intended to set long-distance speed records using a second set of wings with a higher aspect ratio (wingspan) to achieve greater range. On 17 January 1937, he made a nonstop transcontinental flight from Los Angeles to New York in 7 hours and 20 minutes at an average speed of 534 kmph (332 mph).

The Douglas DC-2 was the progenitor of the modern commercial airliner, being of all-metal construction and having multiple engines with seating for 14 passengers inside the cabin. Fitted with retractable landing gear, the DC-2 still left a part of the tires extended below the engine nacelle when the gear was fully retracted to cushion a belly landing due to either mechanical problems or pilot error in not extending the gear.

The single-engine H-1 racer was designed and built for Howard Hughes to set a new world speed record for land planes in 1935. The aircraft reached a top speed of 352 mph using a 14-cylinder radial engine that produced at most 1,000 hp, versus the engines with more than 2,000 hp used in the Schneider Cup speed racers. The high-speed of this airplane was achieved with relatively low horsepower, a strong indication of Hughes's success in producing an aerodynamically clean, flush-riveted, low-drag airframe. (Wings & Airpower Historical Archive)

Cutaway illustration of the internal structure and unique features of the Ryan NYP (New York-Paris). Essentially a flying fuel tank, the NYP was designed to encompass every needed aeronautical element with a minimal amount of structure, and thus weight. The main fuel tanks were located directly ahead of the cockpit, effectively eliminating any forward visibility, and requiring Lindbergh to use a combination of a periscope and leaning out the open window to see ahead. (Wings & Airpower Historical Archive)

Posing in San Diego at what is now Lindbergh Field, the Ryan NYP is shown as it looked before the transatlantic flight with its original rounded spinner. This spinner cracked on Lindbergh's record-setting cross-country flight to New York, and was replaced at Roosevelt Field, Long Island, with the pointed spinner seen in all the post-Atlantic solo-flight photos. The original cracked spinner is on permanent display along with the airplane at the National Air and Space Museum in Washington, DC. (Wings & Airpower Historical Archive)

the navy airplanes came in third and fourth. On 18 October 1922, the world speed record was raised to 359 kmph (223 mph) by GEN "Billy" Mitchell, Commander of the U.S. Army Air Services. The record was further raised to 381 kmph (237 mph) by 1LT Russell Maughan, who had flown the airplane to victory in the 1922 Pulitzer trophy race.

The official speed record was now obviously approaching 400 kmph and the world did not have long to wait. The navy engaged Curtiss to improve the CR-2, which lost to the army in 1922. The navy finished first and second in the R2C in the 1923 Pulitzer race, and a month later new absolute speed records of 417 and 429 kmph (259 and 266 mph) were set by the two navy pilots who had won the Pulitzer race. They used the new standard of a 3-km straight course rather than 1 km, recognizing the higher speeds that were now to be measured.

The 1925 race was won by the army flying the R-3C with a speed of 401 kmph (249 mph), a world closed-course speed record. This was the final Pulitzer trophy for several reasons. Although by today's standards the government expenditure of approximately $500,000 per year on racing is pocket change, in those days aviation budgets were quite small. The United States still felt safe behind its Atlantic and Pacific barriers and the military use of airplanes was still in its infancy. The army and navy had both benefited from the publicity of the Pulitzer races and congressmen who controlled the purse strings did not hesitate to identify themselves in supporting this new military capability.

The shift in racing competition was to switch to the water once the Pulitzers ended. A new race developed, and not only between U.S. military forces but between countries. The U.S. Navy had entered its racing planes in the Schneider Cup competition, an international competition for seaplanes. The increase in engine power shown by Curtiss engines during the Pulitzer races would soon be trivial compared to the increases that would come during the Schneider Cup races; that competition would have a significant effect on fighter design in the second great world war, which was looming on the horizon.

Schneider Cup Races and the Quest for Higher Speeds

The original reason for creating the Schneider Cup in 1913 was to encourage development of commercial hydroplanes, as they were then known. The reasoning was that it would be safer to operate from and over water rather than solid land as a crash into water would be less lethal than a crash into unyielding ground. As aviation

technology matured there was one obvious advantage of waterborne commercial airplanes: A large, heavy airplane did not have to worry about the length of the runway when operating from a bay or river and could take off with plenty of fuel and payload despite the long takeoff distance.

Initial long-distance flights used seaplanes to span the Atlantic and Pacific oceans in the first successful round-the-world flight of Douglas World Cruisers. The hydroplane's design evolved into two different categories for flotation: a flying boat where the fuselage literally took the shape of a boat hull, and a floatplane that landed on two small boat hulls in place of conventional wheeled landing gear. In the post–World War I quest for higher speeds, the Schneider Cup focus for hydroplanes shifted from a concern for safe landings to an effort to produce aircraft that could fly as fast as their land-based brethren.

The golden era of the Schneider Cup races lasted only from 1925 to 1931. With the withdrawal of the American racers after the congressional ban on military racing, racing honors fell mainly to a national government-funded competition between Great Britain

Charles Lindbergh's solo flight in a modified version of a Ryan M-2 monoplane represented a milestone in aviation history not only for technical reasons, but because of the impact of his stunning achievement on worldwide culture. The duration of the flight was 33 hours and 30 minutes at an average speed of 116 mph, covering a distance of 3,520 miles. After the flight, Lindbergh translated his fame into promoting efforts to further commercial aviation, traveling the country and being mobbed by crowds as shown here. (Wings & Airpower Historical Archive)

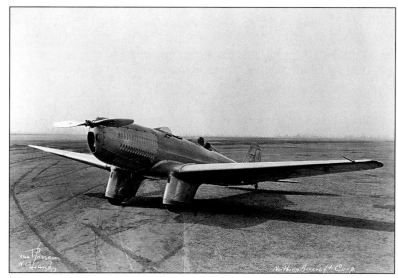

Epitomizing the use of streamlining to reduce drag and increase speed in 1931 was the Northrop Beta, a two-seat single-engine sport landplane. Landing-gear struts and wheels were fully faired over, pilot windshields were highly swept back, and the 160-hp Menasco inline engine's cowling was beautifully faired into the fuselage. The Beta had a top speed of 212 mph, but due to economic conditions during the Great Depression, it was never put into production, and only two were built. (Wings & Airpower Historical Archive)

A record breaker in every sense of the word, Lindbergh's Ryan NYP employed advanced aeronautical design and numerous technical innovations. The valves of the aircraft's 223-hp Wright J-5C Whirlwind radial engine were made of a new high-endurance alloy to ensure continued operation for the 35-hour estimated flight time. A new "Earth Indicator" compass was mounted on the upper aft fuselage, the only available location. Because it was behind him, Lindbergh had to read that instrument with a mirror. The aircraft made 174 flights in nearly 500 hours of flying time. (Wings & Airpower Historical Archive)

Considered the best high-performance production aircraft of its day, the laminated-wood Lockheed Vega was a record breaker right from the start. Amelia Earhart became the second pilot to fly solo across the Atlantic in her red Vega in May 1935. Wiley Post poses here with his world-famous record-setting airplane the Winnie Mae. The powerplant was a 775-hp Pratt & Whitney R-1340 Wasp radial fitted with a new streamlined drag-reducing engine cowling. (Wings & Airpower Historical Archive)

and Italy. As an international sporting competition, the rules said the trophy would be retired if any team won three times. As the speeds increased for the winner each year, it became increasingly difficult as well as expensive to make significant changes to the racer in only one year, so the interval was switched to two years. The Italians and British exchanged victories with increasingly streamlined airplanes with engine horsepower climbing from 800 to 1,900 hp in just three years.

The British Supermarine S5 and S6 designs won twice, but the Italians were hot on their heels and planned a super racer, the Macchi MC 72, for 1931 with twin 12-cylinder engines mounted in tandem producing 2,800 hp and driving contra-rotating propellers. These airplanes were not completed in time for the 1931 race but the British Supermarine S6 was fitted with a 2,300-hp engine and retired the Schneider Cup, also setting a new FAI-recognized world speed record of 409 mph, the first time the record exceeded 400 mph. The Macchi 72 was eventually completed and set a new world speed record for seaplanes of 441 mph, which stood until 1961 when a Soviet jet flying boat captured the prize.

The era of the high-speed seaplane had come to an end, but it indicated what a combination of drag reduction through streamlining as well as more powerful engines could attain. The world speed record now was more than 400 mph and held by a seaplane, which had the drag handicap of carrying two boat hulls to land on. The

One notable example of early long-distance record flights is the little-remembered Russian Tupolev ANT-25 that in 1937 established a new record for distance in a straight line of 6,306 miles. The unpressurized aircraft flew at a maximum altitude of 11,000 feet, and the flight took 62 hours and 15 minutes. More than half the aircraft's takeoff weight was fuel. This crowd must have been amazed to see a giant airplane with red stars on the wings land unannounced in a pasture in California.

12-cylinder engine in the Supermarine S6B eventually morphed into the Rolls-Royce Merlin engine, which was the mainstay for many allied fighter aircraft of World War II. These fighter aircraft achieved speeds of more than 400 mph on the battlefield rather than the race course. The quest for even higher speeds during World War II was approaching a speed limit that the aviation pioneers would never have anticipated, which meant that perhaps an entirely new type of engine would have to be developed.

Altitude and Human Physiology

Human biology evolved with a respiratory system optimized for an atmospheric pressure at sea level of 14.7 psi with the amount of oxygen equal to an atmosphere composed of approximately 21-percent oxygen. At 18,000 feet, air pressure is reduced to half that of sea level and at 33,000 feet, the pressure is one quarter of sea level. At 60,000 feet only one-fourteenth of normal atmospheric pressure at sea level exists.

For example, at 10,000 feet altitude, the amount of oxygen molecules in the atmosphere is lowered to the point that supplemental oxygen is necessary to bring the oxygen level to that at sea level, although it is possible to remain at 10,000 feet for extended periods of time without long-term harm. Physical activity increases the need for oxygen in the body, and for that reason athletes performing in a high-altitude environment often spend a week or two before a competition acclimating their lungs and bodies to reduced oxygen levels.

During World War I, military airplanes operated as high as 10,000 feet and more, but most combats took place lower than that. Toward the latter part of the war, Germans made attempts to provide supplemental oxygen for aircrews by providing an onboard oxygen generator, but the equipment was bulky and dangerous. If a bullet penetrated the apparatus, it caused a conflagration. Other air forces used a small oxygen cylinder with a rubber hose ending in a mouthpiece and a nose clip to provide additional oxygen to pilots for shorter periods of time.

Lack of oxygen at 10,000 feet is generally annoying but not life-threatening. As altitude increases above 20,000 feet, however, so does the danger. Oxygen deficit is clinically known as hypoxia and initially produces behavior resembling alcohol intoxication with higher-order mental functions such as mathematical calculations generally being the first to go. Depending on ambient pressure, lack of oxygen in the blood stream can cause unconsciousness followed by brain death. A figure has been developed known as time of useful consciousness (TUC) and can vary from several minutes at 25,000-foot altitude to 12 seconds in the near vacuum at 85,000 feet where ambient air pressure is less than one psi.

The first high-altitude hypoxia fatalities occurred on the flight of the balloon Zenith to 26,000 feet in 1875. The pilot, Gaston

Pressurized Aircraft, Pressurized Suits

With the invention of the turbo supercharger, it seemed only natural to use a similar installation driven by the engine to provide additional air to the cabin to increase its pressure, eliminating the use of supplemental oxygen. This implied that the cabin was a sealed airtight capsule with air inlets and exhaust valves to allow the pressure to remain at this lower level throughout the flight, no matter what the actual altitude was. Cabin altitude was typically chosen as 5,000 to 8,000 feet and remains that to this day for commercial airliners. In the 1930s, the U.S. Army Air Corps continued its high-altitude research by purchasing and modifying a twin engine Lockheed Super Electra known as the XC-35.

This aircraft was modified with a pressure capsule for the cockpit crew as well as the forward passenger cabin. The size of the windows was decreased from the standard Electra to reduce the risk of a single window failure at high-altitude causing a rapid loss in pressure and decompression sickness for the occupants. A pressure regulation system was installed to keep the cabin at 9 PSI (equivalent to an 8,000-foot altitude) once the aircraft was flying above 8,000 feet. Only one XC-35 was built, but it was so successful that once the test program was over, it was used to ferry general officers around the country. Its success inspired Boeing to design the Model 307 Stratoliner using the B-17's wing, but with a pressurized cigar-shape fuselage. Introduced prior to the United States' entry into World War II, only a dozen were built due to the war, although one was still actively flying in Southeast Asia as late as the mid-1960s.

For high-altitude aviators who were not in pressurized cabins because of their increased complexity and weight, some sort of high-altitude protection suit was necessary. Airplanes could reach altitudes approaching 30,000 feet for extended periods in the 1930s with even higher altitudes possible. The first need was to provide oxygen, but this was becoming increasingly common. Reduced pressure on the human body for extended periods was a different problem. Initial "pressure suits" as they were called were patterned after deep-sea-diving suits and were initially used by Wiley Post in his attempt to set U.S. coast-to-coast speed records by flying above 30,000 feet. Because of the length of these flights he decided that a pressure suit would be desirable for his equally famous supercharged Lockheed Vega named *Winnie Mae*.

Post designed a custom-made pressure suit in cooperation with the BFGoodrich Tire and Rubber Company. The suit was inflated to two psi using air from the engine's supercharger (see page 30), although it was later redesigned to use oxygen from a separate liquid oxygen sphere independent of the engine. He closed the helmet faceplate and pressurized the suit at an altitude of 17,000 feet, and while using it, he actually attained altitudes of 50,000 feet during his long-distance flights. His first attempt at this flight ended with an engine failure and a dead-stick landing at Muroc, California, the future location of Edwards Air Force Base.

He silently glided to a safe landing on Rogers Dry Lakebed but required help getting out of his pressure suit which, without

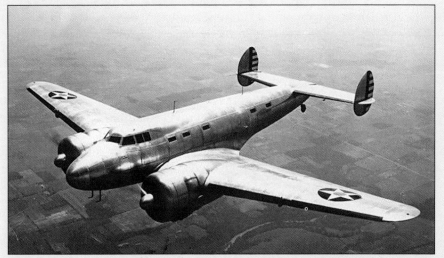

During the 1930s the U.S. Army Air Corps experimented with developing a pressurized cabin in a modified Lockheed Electra designated XC-35. It used air from its superchargers to reduce the cabin altitude to 8,000 feet at much higher altitudes. Window sizes were reduced to diminish the violence of loss of cabin pressure in the event a window failed. The passenger cabin and flight deck were enclosed in a metal capsule reminiscent of a submarine with a pressure door designed to keep air pressure in, whereas a submarine tries to keep water out. (USAF)

The XC-35's success inspired Boeing to design its own pressurized aircraft, the model 307 Stratoliner, which was the world's first operational pressurized airliner. Only a dozen were built for Pan American World Airways and TWA, which both operated them for a short period during the late 1930s. One Stratoliner was still actively flying in Southeast Asia for the International Control Commission as late as the 1960s, however. (Wings & Airpower Historical Archive)

a personal cooling system, was rapidly becoming a heat trap. He attempted to enlist the assistance of a nearby farmer, but when he tapped him on the shoulder, the startled farmer assumed the helmeted figure was an alien from Mars and beat a hasty retreat, which continued until he looked back and recognized the famous *Winnie Mae* sitting on the lakebed. Despite these occasional initial setbacks, technology was developing to set the pattern for future commercial aviation; flying higher, faster, and over longer distances.

It was obvious that better pressure suits than Wiley Post's custom-fitted outfit were necessary. With the advent of jet aircraft, the military took the initiative to develop standard suits to prevent this. First-generation pressure suits of the 1950s were known as partial pressure suits and used tight-fitting fabric for custom-fitted suits to apply 3½-psi mechanical pressure to the human body by inflating small tubes and bladders, or capstans, to make a tight suit even tighter. Early versions of the Lockheed U-2 reconnaissance aircraft used this type of suit, and although effective in an emergency, it was very uncomfortable even when the airplane was still pressurized.

A large improvement came in the 1960s when the decision was made to use pneumatic rather than mechanical pressure. This required building a suit, which was in essence its own capsule sealed to retain air. These suits were much more comfortable and were first used in the Mach-6 X-15 and Mach-3 Lockheed Blackbird, and are still used today in modern U-2s. Retired Air Force Col. Lewis Setter, who flew early model U-2s for years, verified how uncomfortable the partial pressure suits were and stated that he routinely lost 5 pounds during an eight-hour flight.

This retouched composite photo depicts Wiley Post's famed Lockheed Vega 5C, the Winnie Mae, flying at high altitude during a supposed test of his prototype pressure suit. Note the absence of the aircraft's landing gear, designed to drop away on takeoff, leaving the airplane to land on the single skid integrally mounted below its forward fuselage. (Wings & Airpower Historical Archive)

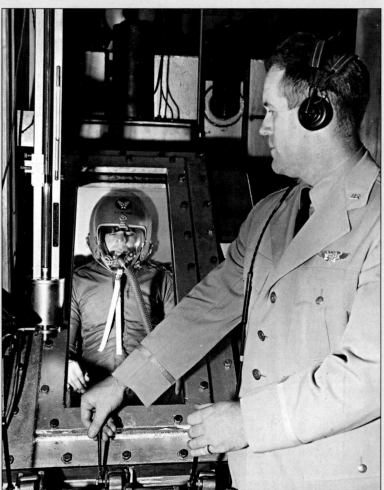

Aviation record-setter Wiley Post intended to fly at high altitudes for coast-to-coast speed records using jet stream winds, but the Winnie Mae did not have a pressurized cabin. Post worked with BFGoodrich to develop the first pressure suit to provide a low-altitude environment for the human body in an unpressurized aircraft. The first attempt ended in failure and Post had to be cut out of the tight-fitting suit, thus destroying it. Later suits were more successful and he was able to verify that winds at altitudes of 25,000 feet and higher significantly increased an aircraft's speed when flying eastward from California to New York.

Jet pilot undergoes training in 1952, using an altitude chamber to simulate the environment in which a pressure suit would be needed in the event of sudden cabin de-pressurization. Capt. James Kennedy (right), an Air Force flight surgeon and pilot, monitors the test subject's performance and operates the controls of this chamber at the Air Research and Development Command (ARDC) aero-medical laboratory at the Wright Air Development Center near Dayton, Ohio. The white lanyard under the chin is used to combat the helmet's tendency to rise as the suit is pressurized.

First-generation pressure suits of the 1950s were known as partial pressure suits, and used tight-fitting fabric for these custom-fitted suits to apply mechanical pressure to the human body by inflating small tubes and bladders to make the tight suit even tighter. Pilots flying early versions of the high-altitude Lockheed U-2 spy-plane used this type of suit, which was effective but very uncomfortable even when the airplane was still pressurized. Here, Maj. Arthur "Kit" Murray wears a T-1 suit and poses next to the Douglas X-3 Stiletto although he never flew that airplane. (Wings & Airpower Historical Archive)

A large improvement came about in the 1960s when the decision was made to use pneumatic rather than mechanical pressure, which required building a suit which was, in essence, its own pressure capsule sealed to retain air. These suits were much more comfortable and were used in the hypersonic North American X-15 shown here after landing, with Capt. Joe H. Engle, who was the youngest pilot to have ever flown that aircraft. (Wings & Airpower Historical Archive)

Tissandier, and two companions had supplemental oxygen in rubber bags, but due to the intoxication of hypoxia, apparently they did not use it. All three lost consciousness and only the pilot survived.

At 25,000 feet, outside air pressure is reduced to one third that at sea level and further problems, known as the bends, can develop in the human body. When breathing normally, nitrogen is dissolved in the bloodstream as well as oxygen. If pressure is dropped too rapidly, the nitrogen forms bubbles, similar to a carbonated drink bottle being opened after being shaken. These bubbles congregate in joints, such as the knees, elbows, and shoulders, and are extremely painful. A preventative measure is called "pre-breathing," or breathing only pure oxygen for an extended period before operating in a low-pressure environment. This flushes the dissolved nitrogen out of the bloodstream. This precaution is used by personnel wearing pressure suits intended for emergency use at high altitudes.

At altitudes above 40,000 feet, the ambient air pressure continues to lower and the human body's ability to process oxygen in the bloodstream is reduced further due to the effect of lower pressure on the lungs. The lower the ambient pressure, the lower the boiling point where liquid turns to vapor, and at 63,000 feet, water boils at human body temperature. This is known as the Armstrong limit (named for a doctor, not Neil Armstrong), and at that point, the lungs essentially fill with steam due to water vapor in the human body, and normal respiration in the lungs ceases. For that reason, flight at altitudes above 50,000 feet requires mandatory use of pressure suits, which apply external pressure to the body in the event of cabin pressurization failure.

Climb-to-Altitude Records and the Supercharger

In the 1960s, Anne Welch, a famous British woman glider pilot, when asked how high one should fly in the then-new ultralight hang glider concept, dryly replied "No higher than you'd be afraid to fall from." The first officially recognized altitude record was 82 feet set by Henry Farman in 1908. In early 1910 the record was 1,000 m (3,281 feet) and by the end of 1910, it had reached 10,168 feet. This progression indicated two things: falling out of the airplane was no longer the primary concern; and flight was now possible at altitudes where the lack of oxygen (or "thinness of the air") was becoming a matter of concern for the airplanes' pilots.

As aviation progressed and aircraft flew faster and higher, certain characteristics of that atmosphere began to come into play. Human beings and engines both need a certain amount of oxygen, with the human's use of oxygen being in the bloodstream that goes to the brain, and the engines use of oxygen being in the combustion chamber where power is developed. For the engine, lack of oxygen reduces available power from the engine; for the human being, lack of oxygen produces impaired function, unconsciousness, and eventually death.

The incorporation of an altitude record was originally spurred by the sporting urge to hold a record exceeding that of other competitors. As the airplane evolved into a tool rather than a plaything after World War I, serious attempts were made to reach much higher altitudes. The loss in engine power at higher altitudes proved to be a problem that had to be solved to reach new heights. One solution was to provide higher-pressure air with more oxygen into the engines cylinders even at high altitudes. This was known as supercharging.

The most well-known effort was started in 1918 by Stanford Moss, an engineer for General Electric. Moss proposed using a turbine mounted next to the engine to compress the air before its entry into the engine's air intake manifold. An American Liberty aircraft engine with a 400-hp sea level rating was taken to the top of Pikes Peak in Colorado (installed in a truck) at an altitude of 14,000 feet, and tests showed that by using the supercharger, engine horsepower could be increased by 20 percent.

Later design developments drove the compressor by using the exhaust gas of the engine to drive the turbine, which then drove the compressor, rather than the original gears attached to the engine crankshaft. Technically this became known as turbocharging. Some aircraft engines employed both supercharging and turbocharging, notably high-altitude Boeing B-17 Super Fortresses, but the supercharging concept itself solved the problem of aircraft engine power loss at higher altitudes.

As early as 1920, the U.S. Army Air Services became interested in high-altitude research when CPT Rudolph Schroeder raised the world altitude record to 33,113 feet over Dayton, Ohio, flying a French-designed Packard Lepere Lusac 11 open-cockpit biplane with a supercharged Liberty engine. An indication of the success of the supercharger was that the service ceiling of the normal unmodified airplane was 20,200 feet. An indication of the difficulties of operating an airplane at these altitudes was that he exhausted his oxygen supply, the reduced air pressure eventually caused the fuel tanks of the aircraft to collapse, and he was so numb with cold that despite wearing a fur-lined electrically heated flying suit, he suffered severe frostbite that actually froze his eyelids open.

Schroeder eventually lost consciousness at 30,000 feet, but fortunately with the airplane already in a descent, he regained consciousness in the lower atmosphere just before crashing into the ground. CPT Schroeder handed over the task of higher records to LT John Macready, and in September of 1921, Macready raised the altitude record to 34,508 feet flying the same airplane, presumably with more oxygen and a modified cockpit heating arrangement. Not to be outdone by the army, navy pilots raised the world altitude record to 43,166 feet, in 1929. The success of the supercharger concept solved a major problem, allowing an engine's horsepower to be retained at higher altitudes.

With supercharged aircraft engines, flight at high altitudes became more attractive for several reasons. Low air density above 25,000 feet meant that drag was reduced and higher speeds were achieved with less powerful and more reliable engines. Balloonists

The Breguet Range Equation and Long-Distance Flights

The Breguet range equation was developed in the 1920s to allow estimating the range capability of aircraft designs:

$$Range = (V/C) * (L/D) * \ln(Wi/Wf)$$

Where:
V = Velocity in knots
C = Specific fuel consumption in pounds fuel per pounds thrust per hour
L = Lift in pounds
D = Drag in pounds
Ln = Natural logarithm, or loge
Wi = Weight of aircraft at start of cruise in pounds
Wf = Weight of aircraft at end of cruise in pounds

Accordingly, for maximum range the ideal airplane should fly at a high speed with an engine system that gives low specific fuel consumption. The lift/drag ratio should also be high, i.e., the wing should develop a lot of lift with low drag.

Once superchargers for engines and streamlined aircraft designs were developed, new distance records could be set. One notable example that is little remembered today is the Russian Tupolev ANT-25 that, in 1937, established a new record for distance in a straight line of 10,148 km (6,306 miles) from Moscow to San Jacinto, California, via the North Pole. Takeoff of the overloaded aircraft was the riskiest portion of the flight and the fence surrounding the airport had to be removed for the takeoff run. The initial record attempt had to land in Vancouver, Washington, after a flight through bad weather of 63 hours and 16 min, but a later attempt in July 1937 established the new record.

discovered that at higher altitudes wind speeds were considerably higher than at sea level, which could be used to shorten journey times if there was a tailwind. Moreover, the prevalence of turbulence (or "air pockets" as it was commonly known) decreased at higher altitudes, leading to a smoother ride with greater comfort for passengers as well as less stress on the airplane. Weather systems within the United States topped-out between 15,000 and 20,000 feet, so flying above the weather was less dangerous. For airplanes to become a serious practical means of transportation, it seemed that higher altitudes were the path to follow.

Pioneering Record Flights and the Traveling Public

Not until after the Roaring Twenties did aviation technology mature sufficiently to make nonstop Atlantic crossings somewhat less hazardous. Engine problems still haunted the long-distance aviator, but by the mid-1920s engines with 400 to 600 hp were available with sufficient reliability for the more than 20-hour Atlantic crossing. The most famous transatlantic flight was that of Charles Lindbergh, or "The Lone Eagle," as he was popularly known. As an airmail pilot who flew single-engine biplanes, he was aware of the increased reliability of engines and estimated that a single-engine monoplane with a single pilot had the best chance of performing a successful nonstop crossing and winning the Ortega prize of $25,000 for the first successful flight from New York to Paris.

The heavily loaded airplane only made this one flight with full fuel tanks, and Lindbergh narrowly missed hitting the telephone lines bordering Roosevelt Field, Long Island. Nevertheless, he successfully navigated across the Atlantic to Ireland, then to England, and then on to France at an average altitude less than 2,000 feet, narrowly

avoiding disaster when he dozed off at low altitude and nearly flew into the ocean. The flight's duration was 33 hours and 30 minutes at an average speed of 116 mph covering a total distance of 3,520 nautical miles. Further records followed over even greater distances (U.S. to Hawaii, Pacific Island hopping, round-the-world flights, England to Australia, Australia to New Zealand, etc.) but Lindbergh's epic flight was inarguably the most famous single flight in history.

The 1930s produced a new generation of more efficient conventional airplanes theoretically capable of carrying people and cargo at higher speeds and higher altitudes. Initially, engine reliability was such that single-engine airplanes were the norm for commercial aircraft such as the Lockheed Vega and Northrop Alpha and Gamma designs, the rationale being that with multiple engines, the probability of losing an engine, and hence losing significant power, was doubled or tripled. For long distances across the considerably larger Pacific Ocean, it was usually necessary to have several engines to carry enough fuel for the longer flight.

The success of the new trends toward metal twin-engine airplanes with retractable landing gear was illustrated in 1935 with the McRobertson long-distance race from England to Australia. Wiley Post intended to win the race by flying the *Winnie Mae* high in the stratosphere, taking advantage of higher speeds and jet stream winds, but was unable to complete the development of the technical details in time for the race. The race was instead won by a special ultra-streamlined twin-engine racer, the de Havilland 88 Comet built specifically for the race. Second and third place, however, were won by almost a standard KLM Douglas DC-2 and Boeing 247 aircraft, which replaced some passenger seats with additional fuel tanks so they could remain in the air longer and with fewer ground stops than other, smaller contestants.

Chapter 3

DAWN OF THE JET AGE (1940–1945)

Perhaps not the sleekest American jet airplane ever built, Bell's pioneering XP-59A Airacomet was definitely the first, making its inaugural flight in October 1942. Seen here flying over Muroc Flight Test Center (now Edwards Air Force Base), the XP-59A prototype is wearing Army Air Forces OD camouflage with gray undersides typical of that era. (Wings & Airpower Historical Archive)

As World War II progressed, fighter aircraft engines became even more powerful with horsepower ranging from 1,600 to 2,000. This gave fighters greater speed and faster climb rates. In air-to-air combat, aircraft with an altitude advantage could translate energy into diving speed and dominate the fight. Aerial combat that started at high altitude was then driven lower as that combat progressed. However, at high altitudes, air is quite cold and airplanes and propellers started to experience a problem that had never been noticed previously.

Flying High-Speed Airplanes with No Propellers

The speed of sound is mainly a function of temperature. Propeller tips on World War II fighters were actually approaching the speed of sound at high altitudes; in fact, the aircraft themselves approached the speed of sound as they dove down from the colder, higher altitudes. Propellers started to lose their efficiency as prop tips became supersonic, decreasing net propulsive thrust. Thus, it seemed there was a basic speed limit for piston-powered fighters, and at speeds above 400 to 450 mph at high altitudes, something other than a propeller would be needed to provide thrust for the aircraft. Fortunately, visionaries in different countries had been studying a new technology that would avoid the supersonic propeller conundrum.

As early as the 1900s, the idea occurred to some aviation inventors to propel airplanes by forcing air out the back of the airplane, pushing the aircraft rather than using a propeller to pull it through the air. In 1910, Romanian inventor Henri Coanda exhibited an airplane at the Paris Air Show using that principle, but the plane was so underpowered it never flew. Twenty years later, two men on politically opposite sides during World War II independently developed the idea for what became the turbojet engine. These men were Frank Whittle in the United Kingdom and Dr. Hans von Ohain in Germany.

They both devised the idea of using a rotary compressor to increase the pressure of the air inflow caused by an airplane's motion in flight, then injecting fuel into the compressed air and igniting it. The resultant hot air then exited at very high speed through a turbine, which drove the compressor (starting the cycle all over again) and provided thrust to push the airplane to higher speeds.

Whittle, a pilot in the Royal Air Force (RAF), first came up with the idea in 1928 while writing a thesis at the RAF College at Cranwell speculating about airplanes of the future. He envisioned air-

With General Electric personnel giving a good sense of scale, this GE 1-A engine is prepared for delivery. Producing 1,250 pounds of thrust, this centrifugal reverse-flow turbojet first took flight powering the Bell XP-59. The GE 1-A was a license-built evolution of the original British Whittle jet engine, and had a time-between-overhaul of three hours. (Wings & Airpower Historical Archive)

craft capable of reaching speeds of 500 mph and decided propellers would not provide the needed thrust to reach those speeds. He envisioned using rocket engines, but realized that rockets were very fuel inefficient, so he devised the turbojet, which used fuel much more efficiently.

The Air Ministry was not particularly interested in his ideas, so he scraped together money (while serving as an RAF officer earning meager wages) to patent his turbojet engine design in 1930. Initially his patent lapsed in 1935 and he could not afford to renew it, but other retired RAF officers not only bankrolled his patent renewal, but also persuaded private companies to invest 50,000 pounds Sterling to develop a working prototype of his design. A woman who was a good friend of his mother, also became interested and invested in the turbojet as well as a new company to be named Power Jet Limited.

Subcontractor BTH was engaged to build the new engine, and soon discovered the challenging technical difficulties of building a centrifugal compressor capable of increasing pressure enough to support combustion, as well as a turbine that could withstand the high temperatures of gases much hotter than a stove. The Air Ministry

Ernst Heinkel and the Race for the World Speed Record

Heinkel was always interested in his aircraft's speed. The Heinkel 70 was in some ways a single-engine Learjet of the 1930s with its elliptical wing and clean lines. It was a favorite for European airlines, and even the British bought one to use as a testbed for advanced engine and propeller designs. Heinkel was aware of the propeller's limitations, but envisioned future airplanes with cruise speeds of 900 or maybe even 1,000 kmph (600 mph). He submitted a design for a monoplane fighter known as the Heinkel 100, which competed with the Messerschmitt 109 design to supply the Luftwaffe for the upcoming war.

In 1938, the Heinkel He-100 with its 1,100-hp engine held the international record for speed over a 100-km closed circuit of 635 kmph (394 mph). On 30 March 1939, a more powerful, 1,800-hp engine allowed the He-100 to hold the absolute speed record of 746 kmph (463 mph). That record only lasted three weeks, however, as a Messerschmitt Me-209 built by Heinkel's archrival set a new absolute speed record of 469 mph. By supporting research into means of advanced propulsion, Heinkel hoped to persuade the Luftwaffe that his company was the correct choice for high-speed fighter aircraft, and for that reason, bankrolled both the Heinkel 176 rocket-powered airplane and Heinkel 178 jet-powered aircraft.

The comparison between the He-100 and Me-209 illustrates the graphic differences in high-speed airplanes driven by propellers. Both were small airplanes but the He-100 was intended to go into production as a fighter. The Me-209, on the other hand, was not a practical airplane and was designed only to set the absolute world speed record, while the He-100's speed record was used for German propaganda to illustrate the superiority of Germany's fighter aircraft.

The Me-209's tiny wing area allowed for higher top speed, the trade-off being a huge, heavy 2,300-hp engine and very limited endurance. The experienced Messerschmitt test pilot who set the speed record described the diminutive aircraft as "a monstrosity, a vicious little brute." He listed 13 factors considered unacceptable for an operational fighter aircraft due to its small size, small wings, and difficulty controlling inflight as well as on landing.

Those phrases echoed Jimmy Doolittle's thoughts when he piloted the Granville brothers' stubby Gee Bee R-1 Super Sportster to a landplane speed record of 294 mph seven years earlier. Ironically, the outbreak of World War II put an end to publicizing exact capabilities of military aircraft for security reasons. The next absolute speed record wasn't set until 1945 and that airplane used a turbojet engine rather than a piston power plant and propeller.

The best of classic Art Deco design can be seen in the elegant lines of the Heinkel He-70, designed for Lufthansa German airlines to compete with Swissair purchasing Lockheed Altairs for high-speed travel between cities in Europe. The Heinkel 70 was faster due to better streamlining and its liquid-cooled engine. The four or five passengers were housed amidships in a relatively luxurious cabin while the pilot sat atop the fuselage somewhat like the driver on a horse-drawn Hansom cab.

finally became interested in the turbojet, classifying the entire project as secret. BTH personnel were somewhat in the dark as to exactly what Power Jet Ltd. was doing; they decided it was probably a new kind of flamethrower, although an incredibly noisy one. During the first ground run in April 1937, the engine over-sped as RPM kept increasing, despite Whittle's best efforts to shut it down.

The noise was incredible and the combustion chamber actually glowed red-hot as if it was about to explode. The rest of the employees all ran away, but Whittle remained in an attempt to control his monster, later admitting he was frozen with fear and could not get his legs to move. Eventually the engine ran down and shut itself off

The Heinkel He-178 resembled its rocket-powered He-176 predecessor, and became the world's first jet aircraft to fly on 27 August 1939. Due to its small size and limited fuel capacity, its first flight only lasted about 5 minutes, and acceleration during takeoff was less than thrilling. The jet was powered by a Heinkel He S-3b engine that produced all of 1,100 pounds of static thrust.

after having escalated from the planned 2,000 rpm to 8,000 rpm. The BTH factory served an eviction notice after this, and Whittle was forced to move his development effort to an unused BTH factory near Coventry. By May 1937 the engine had achieved 13,000 rpm, at which point the turbine disintegrated. While building a new turbine wheel in June 1939, the engine was modified so that speed increased to 16,000 rpm, yet the turbine stayed in one piece. This engine was now known as the W1.

That month the Air Ministry finally signed a contract to build an airplane for this jet engine. Known as the E 28/39, this aircraft was intended strictly as a testbed and was not an operational jet fighter. The contract was awarded to the Gloster Company, whose last production contract was for a biplane fighter called the Gladiator. Since the W1 only produced 855 pounds of thrust (the engine itself weighed 623 pounds!) the airplane was quite small and weighed only 2,800 pounds to accommodate the low available thrust. Frank Whittle was finally going to see his turbojet take to the air, but it was now summer 1939, and the war everyone expected was about to start sooner than anyone had thought.

Summer 1939 was destined to be the last months of peacetime for six years to come. Across the English Channel, another country was also busy investigating the advantages of a turbojet engine. Dr. Hans von Ohain, the man responsible for developing Germany's jet engine, was having an easier time acquiring funding than was Whittle. The sponsoring venture capitalist was a man named Ernst Heinkel, an accomplished aircraft builder in his own right and a visionary looking toward the future.

Hans von Ohain as a young man had already demonstrated his brilliance. He entered the University of Gottingen at age 19 and began his studies to become a physicist, earning his PhD in four years instead of the usual seven. He had no particular interest in aviation

The Heinkel He-280 was a twin-engine turbojet aircraft intended to become an operational fighter. Like many Heinkel designs, it was noted for its aerodynamic streamlining and incorporated a number of features later used by the Me-262, including tricycle landing gear. The jet was equipped with an ejection seat and the pilot of one of its prototypes actually used the seat for the first time in aviation history, and definitely for the first time in a jet aircraft.

The Arado Ar-234 was technically the world's first jet bomber. Equipped with two engines and, eventually four, the Ar-234 was the bomber that the Messerschmitt Me-262 could not be. Unfortunately the single pilot also had to serve as the bombardier, and trust the autopilot to fly the airplane while he was trying to strike a target.

The Messerschmitt Me-262 did see combat against Allied fighters during World War II, but it arrived on the scene too late to make a major difference in the air war over Germany. Shortage of engines and fuel meant that many Me-262s never flew, and the ones that did were so heavily outnumbered that their contribution to the German defense effort was minuscule. (Mike Machat)

at the time, but enjoyed fast cars and partying. As a doctoral teaching assistant at the university, he realized he needed additional income to continue his lifestyle and decided to become an inventor as his second job. He became interested in the idea of a gas turbine engine that could power an automobile or an airplane, and believed royalties from this invention would nicely supplement his income.

One of his professors provided a letter of introduction to aircraft manufacturer Ernst Heinkel who lived not far from the university and was noted for forward thinking and aggressive business practices. In March 1936, von Ohain met Heinkel for the first time and described the principles of the engine, for which he had developed a crude working model. By coincidence, a year earlier Heinkel had met with another young brilliant engineer by the name of Werner von Braun to discuss the concept of rocket engines propelling high-speed aircraft. Thus, the concept of using hot exhaust thrust to propel an airplane was not entirely new to the manufacturer. After discussions with his engineering staff, Heinkel hired von Ohain and his mechanic assistant on the spot, giving them a workshop at the Heinkel factory to build a full-scale engine to propel an aircraft.

Heinkel's backing of advanced engine technology first bore fruit in the form of his He-176 rocket plane, which made its first flight on 20 June 1939. Rather than using von Braun's liquid-fuel rocket engine, Dr. Walther's rocket motor was considered to be more reliable. Again the airplane testbed with the new rocket engine was quite small, with a wingspan of only 5 m (16 feet) and empty weight of only 900 kg (1,900 pounds). It was fortunate the airplane was so small because the Walther rocket motor produced only 1,300 pounds of thrust.

This power proved to be more than enough as the test pilot reported on its first flight that speed increased dramatically from 500 to 750 kmph (310 to 466 mph) immediately after takeoff. Acceleration was so great that his head was pushed back against the seat's headrest. Just a slight move on the stick and the bird shot into the skies at a 45-degree angle without slowing, which gave him a healthy respect for the plane. There was not much time to enjoy the flight, however, as the engine ran out of fuel within a minute after takeoff, emphasizing a problem of rocket versus turbojet. The pilot had to make a very tight turn to set up for diving toward the runway at 500 kmph (310 mph) and barely made it back on the ground, bouncing several times after touchdown.

The following day Luftwaffe generals witnessed the rocket flying, although they were disappointed when they realized the airplane was so small and the rocket burn time so short. General Udet dictated that such a rocket "with running board wings" should not be flown again, as it looked too dangerous. Ernst Heinkel prevailed, however, and it flew several times reaching a speed of up to 850 kmph (528 mph) in a demonstration for Adolf Hitler on 3 July. When the war began on 1 September, all non-essential experimental flying was canceled, and the He-176 ended up as an exhibit in a Berlin museum where it was destroyed in a bombing raid in 1945.

Meanwhile von Ohain and the Heinkel team continued working on the turbojet for the He-178. In spring 1938, a breakthrough allowed an engine to be produced that could generate almost 1,000 pounds thrust. It was successfully flown suspended beneath the fuselage of a Heinkel 118 testbed aircraft. This airborne operation was extremely useful as they discovered there were still problems associated with starting the jet engine, which had to be pre-heated with a hydrogen flame before switching to the kerosene liquid fuel. These flight tests produced improvements and by 27 August the engine was installed in the He-178, ready for the first flight of an airplane powered only by a turbojet. The same pilot who had made the first flight of the He-176 rocket plane was now in the cockpit of the Heinkel 178.

The first thing he noted was that the airplane accelerated more slowly than the rocket, which actually made him feel more comfortable, especially realizing he now had more than only one minute of fuel. After rolling about 900 feet, the airplane suddenly shot ahead and was able to lift off easily. He'd been instructed not to exceed 600 kmph (372 mph) so he made two leisurely circuits around the airfield, admiring the view and listening to the turbine "singing its loud monotonous song," as he put it. After flying for 6 minutes, he began his approach to land, but noticed he was again low on fuel.

Because the airfield was short, he did not want to risk a go-around, so he side-slipped the one-of-a-kind prototype to a safe short-field touchdown. The first turbojet flight was now history. It was not until 1 November that General Udet was able to watch a jet flight again, but although he complimented the pilot on his successful flight, he wasn't overly enthused with the small jet testbed's performance. Nevertheless, he promised Heinkel that if he could produce a flyable jet fighter by April 1941, he was awarded a production contract for the jet fighter.

Jet Fighter Competition: Heinkel versus Messerschmitt

Heinkel recognized that his biggest rival in producing fighters for the Luftwaffe was the company run by Willy Messerschmitt that won the contract for the Me-109 fighter. In his forward-thinking entrepreneurial style, he hired away one of Messerschmitt's best designers, Dr. Robert Lusser. Lusser was the man who designed the Messerschmitt 110 twin-engine fighter and he seemed a natural to design the jet, which also required twin engines.

The design became known as the Heinkel 280 and incorporated several advanced features in addition to the jet engine, including

---

Operating Principles of Turbojet and Rocket Engines

To compare engines in common conditions, the thrust rating of a jet engine is usually given as static sea level thrust where V1 equals zero and the mass flow of the working fluid (air) assumes standard air density at sea level.

$$\text{Thrust (pounds force)} = Q\ (V2\text{-}V1)$$

Where:
Q = Mass flow through the engine (slugs per second)
V1 = Velocity of air entering the engine (feet per second)
V2 = Velocity of air exiting the engine (feet per second)

In the case of the turbojet, air approaches the engine at velocity V1 depending on airspeed. Within the turbojet engine the air is compressed to a higher pressure by rotating high-speed machinery. Energy is added to this airflow by mixing and burning fuel in the combustion chambers. The resultant high-speed mass flow is expelled from an exit nozzle finally reaching velocity V2.

By comparison, a rocket power plant produces thrust by creating a very large change in velocity of a relatively small mass of fuel and oxidizer that are both housed within the vehicle itself. Because the rocket is completely self-contained, its thrust is independent of the speed of the engine and is actually equal to the mass flow times the exhaust velocity. Rocket exhaust velocities are generally very high to produce large amounts of thrust, so rocket engines are generally inefficient because they must carry their own atmosphere with them for combustion, which quickly exhausts their fuel supply.

Turbojet engines are best at high speeds where piston-powered aircraft have problems. Best speeds occur at high RPM at high altitude. At low altitude at high speeds, the jet may produce more thrust, but the drag force is also higher. When speed records were allowed to be set at high altitudes versus only 100 meters, huge jumps in speed were achieved. Jet engine mass flow is sensitive to air density, which is a function of altitude and temperature, so takeoffs on a hot day can sometimes be lengthy due to low thrust buildup during the takeoff roll, leading to increased runway distance used.

tricycle landing gear. This was an important feature because the two jet engines blew very hot exhaust gases well behind them. A conventional tailwheel jet airplane had a tendency to set fire to grass runways or melt asphalt, as the rival Me-262 later proved. The tail section that resembled the Me-110's twin rudder was mounted on a small pylon at the end of the fuselage to ensure that the hot gases did not interfere with the rudders.

In addition, the pilot was equipped with an ejection seat powered by a pneumatic charge intended to get the pilot clear of the airplane in an emergency at high speeds. Heinkel himself was a pilot who taught himself to fly in 1910 and who was badly injured while doing that. Throughout his career he maintained a strong interest in pilot survival. Design of the Heinkel 280 progressed swiftly and the aircraft made its first flight on 22 September 1940, unfortunately as a glider, because the engines were still not ready.

The centrifugal flow He58 engines each produced 1,650 pounds of thrust for a total of more than 3,000 pounds thrust. The airplane had a wingspan of 40 feet, was 34 feet long, and weighed 6,700 pounds empty, fairly reasonable for a fighter equipped with cannon armament. Jet engines were still quite rudimentary, which is why so many early designs had to have twin engines to power them. New, higher-thrust engines had some of the same problems as the early Heinkel engines and initially leaked so badly that for the initial flight, the pilot insisted on leaving the engine cowlings off. This wasn't to cool the engines, but rather to keep fuel from puddling in the bottom of the nacelles, leading to possible fires.

That first flight took place on 30 March 1941 and lasted only 3 minutes, but it showed that the design was sound while at the same time beating General Udet's April 1941 deadline for a production contract. The celebratory parties were once again historic. A few days later on 5 April, a demonstration flight complete with closed cowlings was made for General Udet, who kept his word and eventually awarded a production contract for 300 aircraft.

In Germany, Messerschmitt was working on prototypes of an operational jet fighter. Like the Heinkel He-280, which had already flown, the Messerschmitt 262 was powered by twin turbojets. These engines were different from Heinkel's because they employed axial flow compressors, unlike earlier jet engines, which used centrifugal flow. Axial-flow engines employed several stages of compression to produce more thrust and were narrower than centrifugal flow engines, producing more net thrust due to less frontal-area drag. The Me-262's engines were built by the Jumo Corporation and were designated Jumo 004. The two engines provided almost 4,000 pounds of thrust total, and propelled the aircraft to 500 mph at sea level; slightly higher at altitude. Takeoff weight was 14,700 pounds and it was equipped with four 30mm cannons later supplemented by unguided rockets.

Like the He-280, the Me-262 was ready before its turbojet engines. The first flight was made on 18 April 1941 powered by a single piston engine with a propeller. It flew well as a propeller driven

This rare photo of the Gloster E.28 demonstrates the fact that the size and width of the turbojet engine drove the design of the airplane. With an air inlet in the nose and an exhaust in the tail, the engine had to be in the middle of the airplane's fuselage for center of gravity reasons making the aircraft rather portly, since engine diameter was considerable. (Wings & Airpower Historical Archive)

airplane. On 25 March 1942, the prototype took to the air with BMW axial flow turbojets, but the piston engine was kept as an insurance policy. This turned out to be a wise decision, as both turbojets quit due to compressor blade damage suffered while taking off on the unpaved airstrip. The pilot started the piston engine and landed successfully but incurred landing gear damage because the aircraft was still heavy due to the extremely short flight.

On 18 July 1942 pilot Fritz Wendel tested the prototype using the Jumo jet engines without the piston engine and prop. A longer paved runway was used, but the pilot was surprised to discover that as he pulled back on the stick he could not raise the tail of the aircraft off the ground. (The first Me-262 prototypes were equipped with "taildragger" conventional landing gear.) He remembered somebody mentioning to tap the brakes just before takeoff to allow the tail to rise. That technique worked and he made a 12-minute flight. He was enthusiastic about the airplane and the turbojet engines but had difficulty landing because the idle thrust of the engines was set too high.

On the second flight of the day he reached a speed of more than 400 mph at 10,000 feet. He also discovered that the lack of propeller wash over the elevators was significant on takeoff, and that the taildragger configuration was impractical for jets. This made tricycle landing gear almost mandatory for turbojet airplanes.

The third prototype Me-262 crashed in August 1942, unable to become airborne due to hot weather and reduced engine thrust at high ambient temperatures. Further tests showed that the airflow over the wing center sections separated prematurely at slow speeds, which was one of the reasons for the jet's high landing speeds. In

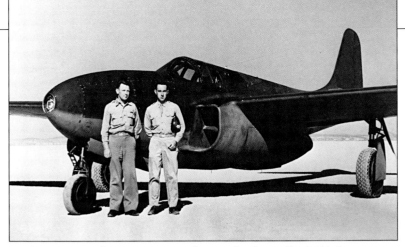

With America's first jet-powered aircraft, the Bell XP-59 Aeracomet as a backdrop, Col. Laurence C. "Bill" Craigie poses with a fellow Army Air Forces officer on Rogers Dry Lake at Muroc in October 1942. It would have been hard to imagine back then that this very spot would someday be the landing site for manned space shuttles returning from a week in orbit. (Tony Landis Collection)

an attempt to correct this problem, the wing center section was extended forward, producing a wing planform that was almost a perfect 19-degree sweep back.

The (British) Empire Strikes Back with a Meteor

Meanwhile back in England, Frank Whittle was still an officer and pilot in the Royal Air Force. The outbreak of war there caused concerns that he'd be mobilized, bringing to an end the turbojet engine development. By now the airplane was being built to house that engine and the Air Ministry decided Whittle would be more valuable left where he was. The Gloster E 28/39 never received an official name, but was informally called "the Squirt" around the factory. Because England's jet engine studies were still officially classified, that nickname may have given away too much information, so there was never any consideration to making the aircraft's name official.

The W1 produced 1,240 pounds of static thrust at 17,750 rpm. The W1X ground version first ran in December 1940 and was superior to earlier pre-prototype engines. Its development proceeded at what seemed like a glacial pace to Whittle, but the factory's location close to the infamous target city of Coventry and the occurrence of the Battle of Britain overhead may have had some effect. Gloster completed the aircraft in late March 1941, but its flight-worthy engine was not ready, so the ground prototype W1X was installed for initial taxi trials.

The first taxi run was made on 6 April 1941 and the test pilot was disappointed to say the least. The airfield grass and recent rainstorms made the ground wet and soggy. In addition, a governor had been installed so the pilot could not advance the RPM beyond 12,000. As a result, the airplane did not taxi faster than 20

mph and the entire Gloster crew was surprised; they'd expected the aircraft to do a high-speed taxi run. Whittle explained to test pilot Jerry Sayer that the turbojet was known to be inefficient at low speeds. Sure enough, the following day the governor was increased to 16,000 rpm and taxi performance improved considerably to the point where Whittle himself took the controls and taxied up to 60 mph.

Whittle later commented that he had realized that with only a slight pull on the stick he could have been airborne, but because the airfield was quite small he might not have been able to stop. Sayer then took over and made two fast runs, lifting-off at 200 or 300 yards each time. Needless to say, he was much happier than the previous day as he'd been concerned the whole effort would be for naught. Several days later the airworthy W1 engine arrived and was assembled and duly installed in the aircraft. On the evening of 15 May 1941 the weather lifted sufficiently for a first flight.

The takeoff run was approximately 600 yards and the new prototype aircraft disappeared into the clouds shortly after takeoff. People on the ground saw little of the airplane after that until 17 minutes later when Sayer made his landing approach at high speed, which indicated to Whittle that the test pilot was happy with his new toy. No film footage was made to commemorate the historic event; the Air Ministry had ignored a request for an official photographer to document it. The only surviving pictures were taken by an unofficial bystander using an unauthorized camera.

For a limited time only in the test program, the Air Ministry gave permission for the RPM governor to be set to 17,000 rpm, providing higher thrust, and at this power setting, the E 28 clearly demonstrated that it outperformed the Spitfire in terms of pure speed. This was illustrated about a year later when the E 28 was flown to another airfield for a demonstration to Prime Minister Winston Churchill, accompanied by an escort of two Spitfires and a Tempest to basically provide protection in the event a curious RAF pilot tried to jump this strange-looking airplane. The Squirt left its escorts far behind, landed, and taxied-in before the embarrassed escorts arrived at full throttle.

Two aircraft were built and one is still preserved in the Museum of Science in London and displayed as Britain's first jet airplane. It had already been decided that a twin-jet fighter should be developed and an initial contract was awarded long before the first flight of the E 28. That aircraft was to be powered by the W2 engine with 1,600 pounds of thrust. This commitment was made even before the Battle of Britain in the summer of 1940, and Whittle dryly commented that he believed the Air Ministry's enthusiasm clouded their judgment.

The success of the E 28 gave the Air Ministry confidence that a twin-jet fighter would be simple to achieve. The fighter was intended as an interceptor because the RAF was quite interested in home defense. Twelve prototypes were ordered, although only eight were actually built. The first flight of prototype G 41, now named Meteor, was made on 5 March 1943 using two 1,500-pound-thrust

Bell's XP-59A was essentially a hurried insertion of jet engines into an airframe not optimized for high speeds. The XP-59A was even larger and heavier than contemporary European fighters, making its thrust-to-weight ratio rather anemic at best. A total of 66 production P-59 airplanes were built, used mainly to familiarize pilots with the handling characteristics of jet aircraft. (USAF)

de Havilland Goblin turbojets. The F1 production Meteor used two Rolls-Royce Wheland engines of 1,700 pounds thrust each. It obtained a maximum speed of 385 mph at sea level and 410 mph at 30,000 feet, which in reality was not much better than the Mustang.

Production rapidly shifted to the Meteor F3 equipped with two Rolls-Royce Derwent engines of 2,000 pounds thrust each, and this increase in power raised maximum speed to 475 mph at 30,000 feet with an initial climb rate of 4,000 feet per minute. More than 200 Meteor F3s were built. A Derwent 5 engine was developed for Meteor with 3,500 pounds of thrust, giving the Meteor F4 a top speed of 585 mph at sea level and a stunning initial climb rate of 7,300 feet per minute.

The twin-engine British jet fighter had been rushed into production in 1943 as a result of intelligence discovery of an impending threat: the V-1 "buzz bomb" (officially named Fiesler Fi-103). Powered by a pulse jet of fairly low thrust, but very simple construction, the pilotless aircraft flew at low altitude at speeds of more than 350 mph to deliver a 2,000-pound warhead somewhere in the English countryside. In June 1944, 2,400 of the weapons were launched, aimed generally at London, and it was obvious that buzz bombs were a real threat to the civilian population. It was determined that even with the low-thrust jet engines of the time, Meteor had a slight speed advantage, and at a typical 5,000-foot altitude, could overtake the V-1 and shoot down the pilotless bombs.

The V-1 campaign started shortly after D-Day, and once Meteors were stationed close enough to the coast, they were able to engage and destroy a total of 13 V-1s. Because of the limited number in service at any given time, usually only two Meteors were airborne together on patrol. Once the buzz bomb threat was over in fall 1944, RAF pilots had an opportunity to take a closer look at the Meteor as a fighter to engage other fighters. In December 1944, they took delivery of the Meteor 3 with its twin 2,000-pound-thrust Derwent engines. The aircraft was still somewhat underpowered, suffered from poor visibility to the sides and rear, and had extremely heavy ailerons, which made dogfighting a real chore.

Engineers deliberately designed the aircraft with very heavy aileron forces because they feared torsional twisting of the outboard wings during high-G maneuvering. None of the pilots noticed it because aerobatics were forbidden in the earlier, even-more-underpowered Meteors. A further problem was that at speeds above Mach 0.67 the aircraft suffered from a directional "snaking," which made it difficult to keep the gunsight on target. (Mach number speed is a percentage of the speed of sound with Mach 1.0 representing the local speed of sound.)

If speed was increased to 510 mph at 5,000 feet (Mach 0.72), violent snaking and lateral oscillation ensued, which required considerable effort to move the stick. At Mach 0.73 the test pilot reported violent aircraft vibration, with the stick vibrating badly and becoming almost impossible to move. Upon throttling the engines back and slowing the aircraft, the controls became effective again after a short pause.

Based upon these reports, the Meteor 3 was restricted to 500 mph indicated airspeed at 6,500 feet and below. Wind tunnel tests at Farnborough later showed that the vibration was caused by airflow separation between the nacelles and the inboard wings. By installing longer nacelles this could be avoided and the limiting Mach number could be increased to Mach 0.84, or 75 mph faster at medium and high altitudes. As the Allied armies advanced into Belgium toward Germany, one squadron of Meteors was deployed to the continent in January 1945 using Meteor 3s with short nacelles.

Their initial problem was that Allied anti-aircraft gunners did not recognize the Meteors and fired at them, fortunately with little effect. It became such a problem that the Meteors were painted all white to be instantly recognizable. German fighters seldom ventured over the Allied armies, and as a result the Meteors never had a chance to engage an Me-262 jet. Seven German aircraft were destroyed by Meteors, but they were all on the ground and were strafed by the jets. Now at last, the Royal Air Force had an operational jet fighter.

Seen here posing on the ramp at Muroc's super-secret North Base complex in the late afternoon sunlight, the XP-59A is readied for flight with Bell test pilot Bob Stanley preparing to enter the cockpit. This photo was taken from the roof of Hangar 1, now preserved at Edwards as a historical landmark. (Wings & Airpower Historical Archive)

Shown parked on the ramp at Wright Field in Dayton, Ohio, this Messerschmitt Me-262 was one of many captured German aircraft brought to the United States for thorough evaluation and study after World War II ended. (Wings & Airpower Historical Archive)

Americans Enter the Jet Engine Game
Thanks to Their British Cousins

General Arnold of the U.S. Army Air Forces was invited to witness a test flight of the Gloster airplane, and realized that with jet propulsion, fighters of the future would be drastically different. He requested blueprints of a sample jet engine be shipped to the United States for study, which the British agreed to. General Electric was chosen as the company to analyze the jet because of their experience with superchargers, while Bell Aircraft at Buffalo, New York, was chosen to build the airframe to give the Americans experience with the jet engine. Called the XP-59 Airacomet, it was secretly tested at Muroc Army Air Force Base, which eventually became Edwards Air Force Base, the Kitty Hawk of the jet age in America.

Due to the low 3,000-pound thrust of the initial GE jet engines plus the large size of the XP-59, the airplane had to be a twin-engine design. The XP-59 made its first flight from the dry lakebed at Muroc in October 1942 with Bell Chief Test Pilot Bob Stanley at the controls. Even with two engines, it could only fly 430 mph at 30,000 feet, 50 mph slower than the P-51 Mustang. Nevertheless, it was America's first jet airplane, and more than 100 P-59A production versions were built, although used more for familiarization than as operational jet fighters.

Today, a silver P-59A is mounted on a pylon in front of the headquarters building at Edwards as a reminder of the United States' proud jet heritage. But with rapid increases in jet engine thrust as the war progressed, an even better jet fighter was envisioned as the United States' first true jet-powered warplane. Development of this aircraft, known as the Lockheed P-80 Shooting Star and completed in only 142 days from contract to roll-out, was a product of an engineering genius named Kelly Johnson. His team, and the super-secret organization that created the airplane, was destined to go down in history.

From Testbed to Weapon:
Luftwaffe Jets Become a Last-Ditch Defense

The initial order for prototype Messerschmitt 262 aircraft guaranteed that it would probably eventually go into production, but it turned out to be a slower process than people originally had thought. Problems with the Jumo 004 jet engines delayed the program by more than a year as the engine had to be redesigned to not use strategic high-temperature metals that were not available for large production. In 1943 operational fighter pilots were introduced to the prototype Me-262s and were amazed. General Adolf Galland, an experienced fighter pilot and top ace, described the feeling, "as if the Angels were pushing me."

Another Adolf (Adolf Hitler), however, had other ideas in mind. After seeing a demonstration, he intended to use the airplane as a blitz bomber rather than as a fighter. Germany was actually designing a true bomber, the Arado 234, which was still in the development

process. No one could convince Hitler of the difficulties of dropping a bomb and hitting a target at the 500 mph speeds of which the Me-262 was capable. Once bomb racks were installed on the Me-262, pilots and crew found that the only way to hit a target was to enter a 15-degree dive with the engines throttled back and release the bomb at low altitude.

Hitler did not approve of this method and other attempts were made, including one using an Me-262 with a glass nose with a bombardier lying flat on his stomach in the gun compartment of the fighter. By 1944 it was obvious that the Me-262 would be far more valuable in defending the Reich against the bomber fleets with their swarms of escort Mustangs than acting as an inaccurate high-speed bomber. The combination of Hitler's insistence on a bomber version combined with the delay in production of Jumo jet engines meant that it was not until October 1944 that the first operational jet fighter unit was organized.

The Me-262's entry into operational service was noted with amazement by the Allied air forces. Initial reports came from high-speed Mosquito and Spitfire photo reconnaissance airplanes at high altitude. They reported being intercepted by a twin-engine propeller-less airplane that was faster than they were and could reach them at the altitudes at which normal propeller driven fighters were struggling to make the intercept. Soon, B-17 bombers and P-51 Mustangs were reporting encountering the same airplanes at lower altitudes flying at speeds that left the Mustangs in the dust.

Occasional dogfights ensued between Mustangs and these new strange airplanes, but usually the results were indecisive, with no

This close-up of the Jumo 004B turbojet mounted to the Me-262 gives a good indication of jet powerplant state of the art in the early 1940s. Thrust-to-weight ratio of this airframe/engine combination allowed unheard-of climb rates and airspeeds as high as 540 mph. (Wings & Airpower Historical Archive)

Captured Heinkel He-162 Salamander, also known as "the People's fighter," makes a test flight from Muroc flown by 1Lt. R.A. "Bob" Hoover. The aircraft was built from nonstrategic materials (mainly wood), had a single Jumo turbojet, and was equipped with an ejection seat due to the jet inlet positioned directly behind the pilot. Although nearly 300 of these unique aircraft were built as a last-ditch effort to thwart growing attacks by Allied bombers, none saw operational service.

one being shot down. B-17 gunners had a very difficult time shooting at an airplane flying 300 mph faster than they were, and a number of B-17s were shot down by the heavy cannon armament in the Me-262s. Countermeasures were soon developed as photo reconnaissance revealed the location of air bases where the jets came home to roost.

At this point in the war, many aircraft were coming off production lines but most did not yet have engines, and for those that did, fuel was in short supply. As a result, although 1,400 Me-262s were built, there were never more than 200 operational and at only a few airbases. Mustangs were assigned to patrol over the airbases where the jets were known to be based. Whenever an Me-262 took off, Mustangs performed high-speed dives on the jet fighter while it was still slow, shooting it down.

Approximately 115 Allied airplanes fell victim to Me-262s, while 100 of the jets were shot down by Allied aircraft, mostly P-51 Mustangs. The Me-262's slow-firing 30mm cannon meant that if they hit a B-17, they would probably bring it down, but the high closure rate made it extremely hard to aim precisely and fire in the short time the target was in range.

In addition to the Me-262, the Luftwaffe had several other jet- and rocket-powered combat aircraft by 1944. The Arado 234 was the first jet bomber and was intended for reconnaissance as well as bombing. It had two or four jet engines and was flown by a very busy single pilot who performed not only as the pilot, but also the navigator, bombardier, and photographer of this high-altitude high-speed aircraft.

Additional jet aircraft used at the end of the war included the Heinkel 162 Salamander, otherwise known as the People's Fighter (Volksjaeger). The He-162 was equipped with a single BMW 003 turbojet engine of 1,750 pounds thrust. Originally intended to be flown by pilots with very little flying experience and virtually none in a jet-powered aircraft, it quickly became obvious to test pilots who flew the little jet that although its performance was outstanding, it could not be flown by inexperienced pilots.

Designed as a desperation fighter to be mass produced in large numbers using non-strategic materials such as wood, the aircraft was designed and built in less than three months. It was intended to carry an armament of two 30mm cannon with a takeoff weight of 5,500 pounds and a speed of 490 mph at sea level; 520 mph at 20,000 feet. The first aircraft flew on 6 December 6 1944, and flew very well. Nevertheless, on its second flight the right wing tip separated from the airplane followed immediately by the entire leading edge of the wing, probably due to inferior glue. The aircraft crashed, killing the pilot. (The He-162 was equipped with an ejection seat using an explosive charge but the pilot never had a chance to use it.) Later aircraft corrected that problem and flew quite well.

The famous English test pilot Eric Brown flew it after the war and was quite impressed with the airplane, referring to it as "a delightful little machine." A total of 116 Salamanders were produced in the short time available but never became operational aircraft, and most of them never flew because the war in Europe ended in May 1945.

Another German jet prototype stirs controversy to this day, the Horton Ho-IX twin-jet flying-wing jet fighter. It was a radical design in aerodynamics being a pure flying wing using elevons and drag rudders for flight controls with no vertical tail. Production versions would have been built by the Gotha Corporation and were designated Gotha 229, but the first prototype never received its jet engines and flew strictly as a glider, as were most earlier Horton designs. The second prototype flew with two Jumo 004 turbojets, but after only two hours of flight testing the plane crashed; the third prototype had not been completed by the time the war ended.

The Gotha 229's estimated performance included a maximum speed of 590 mph at sea level and a service ceiling of 52,000 feet. These figures were never confirmed in flight test, but the uncompleted third prototype was shipped to the United States for study, although it also was never flown. In the late 1980s when Northrop's B-2 stealth bomber was unveiled, some observers noted the resemblance of the Gotha 229 design to the B-2 and interpreted this to mean the German aircraft was a stealth jet fighter. A television program was even produced using a full-scale mockup of the Gotha 229 to examine its radar cross-section. It showed that the flying wing did indeed have a lower radar cross-section than a conventional aircraft, but that may have been a byproduct of the Gotha 229 design rather than a major design objective.

One further example of German advanced propulsion technology entering operational service was the Messerschmitt 163 rocket-plane. Despite turbojet engines being more operationally useful than rockets, the Luftwaffe continued research into rocket-powered airplanes at their secret Peenemünde test base, resulting in the development of a new tailless aircraft. Designed to take a Walters rocket motor, this radical new aircraft was powered by hydrazine/methyl alcohol and highly concentrated hydrogen peroxide, a lethal cocktail that was not only enormously explosive but extremely toxic, and capable of instantly dissolving organic matter, such as the pilot and ground crew.

The first Me-163 flew in April 1941, but did not enter operational service until summer 1944. The aircraft originally flew as a glider in 1940 and later had a propeller with an 85-hp piston engine in the nose. The rocket prototype commenced powered flight in April 1941, and on May 10 test pilot Heintz Dittmar attained a speed of 623 mph in level flight, an unofficial "speed record" actually set during World War II. Empty weight of the Me-163 was 4,200 pounds and it had a 30-foot wingspan with an 18-foot length. Its rocket motor had a thrust range from 500 to 3,750 pounds, although the motor's endurance was only 4 minutes at full thrust and 8 minutes when operated at lower thrust.

As a result of this limitation, the 163 was considered a point defense interceptor with a range of only 70 miles. As an operational weapon it was of limited utility. Its high speed and fantastic climb rates allowed it to escape the escort fighters, but it had the same problem as the Me-262 jet of closing on the slower bombers at more than 400 feet per second and not having time to aim its 30mm cannons. Escort Mustangs soon learned to orbit over the base where the little gliders landed and pick them off there. A total of 364 Me-163s were produced, but it is doubtful that more than 25 airplanes were operational at any given time. They accounted for the destruction of only 16 Allied aircraft, but suffered far higher losses themselves for numerous reasons.

These included limited training for the pilots, and fuel limitations for training sorties because the only factory that made the exotic rocket fuel had been bombed out and it took some time to get back online. Eric Brown, the British test pilot who flew the Me-163 as a glider after the war, commented that the airplane itself was probably a greater danger to its pilots than Allied defenses. Since it landed on a skid like a glider, for takeoff the Me-163 had a two-wheel trolley that was supposed to release after takeoff. However, if there was a problem, it usually led to the fatal crash of a fully loaded airplane. The limited duration of powered flight also put rocket pilots at a tactical disadvantage against the Mustangs.

Overall in a comparison between turbojet airplanes and rockets, as a weapon the turbojet craft proved to be superior. This trend continued for the second half of the 20th Century. Nevertheless, the rocket engine showed what could be done in attaining high-speed flight, and later became the power plant of choice for exploration into the supersonic realm. The development of the turbojet engine meant that in the years following World War II, supersonic flight would probably become a reality; the question now was how to design an airplane that routinely exceeded the speed of sound in level flight. It was obvious that significant new research had to be done to answer this question.

Dramatic ant's-eye view of the Heinkel He-162 at Muroc shows the aircraft's narrow-tread main landing gear and gaping maw of an air intake located above and just behind the cockpit, a feature not seen again until the Mach-2 North American F-107A about 10 years later. Note the downward-canted wingtips to augment lateral stability. The aircraft's right-hand 20mm MG-151 cannon has been removed and its aperture faired over. (Wings & Airpower Historical Archive)

The Messerschmitt Me-163 rocket-propelled fighter interceptor was designed to combat Allied bomber fleets over Germany, but suffered from a number of flaws. As with all rocket-propelled aircraft, it ran out of fuel within several minutes and became a moderate-speed glider. Its rocket engine could propel the aircraft to speeds of up to 600 mph, too fast for the pilot to aim at a target during the limited amount of time he was within gun range. (Mike Machat)

Kelly Johnson and the Skunk Works: Lockheed Develops a True Jet Fighter

Turbojet development was not standing still. Engine technology had increased rapidly to the point where thrust levels were higher and a jet fighter only requiring one engine could actually be flown. For the United States, the Lockheed P-80 Shooting Star was destined to be America's first true operational jet fighter as well as the initial product of what became the most famous "top-secret" aircraft shop in the entire world.

In May 1943, a conference took place in Washington, DC, were discussions were held between the Army Air Forces and Lockheed representatives about progress in jet propulsion. Lockheed was invited to develop an airplane around the new H1 Goblin centrifugal-flow turbojet providing 2,500 pounds thrust. Lockheed's talented Chief Designer Kelly Johnson had actually been studying the prospect of a turbojet-powered fighter using a Lockheed-developed axial-flow turbojet in the early 1940s, but the company never proceeded with that project.

Based on his earlier studies, Johnson felt it would be possible to deliver a prototype jet fighter in record time as long as he was given authority to design and build the aircraft employing a small group of men and unusual streamlined, project-oriented management techniques. This included having design engineers and shop personnel working together to build a prototype in a minimum amount of time with the project engineer having ultimate authority to make decisions. An initial proposal to build one prototype aircraft was generated by Lockheed and accepted by the Army Air Forces within two months.

The contract specified a schedule of 150 days to first flight from contract signature. The project received the Lockheed designation of L140 but the army air forces designation was XP-80. The new organizational structure within Lockheed included only 100 men and specified working 10 hours per day, 6 days per week until task completion. Secrecy was of paramount importance, not necessarily because of the enemy, but because the fewer people who knew, the less interference the project engineer would have. This organization, eventually known as the Skunk Works, produced many advanced technology prototypes in the following decades.

This captured Messerschmitt Me-163 at Wright Field underwent testing and evaluation in powerless glide flights. It is reported that Air Force 1Lt. R. A. "Bob" Hoover made some of the test flights of this aircraft. Note the Me-163's twin mainwheel dolly that drops away at takeoff, and the fixed semi-recessed tailwheel. It was a rocket-powered "taildragger"! (Wings & Airpower Historical Archive)

Lockheed XP–80 "Lulu Belle," designed by Lockheed's Kelly Johnson, was completed in only 143 days from issuance of contract. It was the first aircraft to stem from his secret organization known as the Skunk Works. Despite its dated straight-wing design, the production-version P-80 scored the first jet combat victory, shooting down a more advanced MiG-15 in Korea, and became the basis for the two-seat T-33 jet trainer and F-94C Starfire jet interceptor.

Meanwhile, work on the original XP-80 moved 70 miles north from Burbank to Muroc Army Air Force Base in the high desert at the same isolated North Base complex where the Bell P-59 had first flown. The promised H1 engine finally arrived and was installed in the aircraft, starting on the first attempt with thrust measured at 2,460 pounds. On the second day of ground engine runs, however, the intake ducts on the aircraft collapsed and pieces were sucked into the running engine, seriously damaging it.

On 28 December, a replacement engine arrived and was quickly installed, with ground runs performed on New Year's Eve. After taking the New Year's holiday, taxi tests began, and 8 January was chosen for the first flight. Three busloads of Lockheed personnel (including the entire Skunk Works staff) made the trek from Burbank to Muroc and were in place at the north end of the lake bed at 08:30 in the morning. Test pilot Milo Burcham took off at 09:10, circled once, and landed 6 minutes later. He had been unable to retract the landing gear, so some adjustment was required.

Burcham took off again for a 20-minute flight, which included high-speed low passes and aileron rolls in front of the crowd. Upon landing, he commented that at 490 mph, he had flown the XP-80 faster in level flight than he'd ever dived the P-38. The cockpit was relatively quiet and there was no vibration or buffeting. Maneuverability was excellent and the aircraft's roll rate was very fast and easily controlled. The XP-80's test program continued until 13 April and a top speed of 506 mph in level flight was easily achieved.

Further tests included gun-firing trials and even spin tests performed by Lockheed test pilot Tony LeVier. The XP-80 ended its days as a training aircraft, followed by ground engine tests, and was finally put in storage on 10 June 1946. Eventually it was completely restored to its original condition and proudly put on display at the Smithsonian National Air and Space Museum in downtown Washington, DC, as America's first true jet fighter.

The XP-80A required a fairly complete redesign to accommodate the new higher-thrust jet engine. The aircraft's wing span and fuselage were expanded by 2 feet each, and the airplanes were planned to have pressurized cockpits. Tony LeVier was chosen to make the first flight of the first XP-80A (called *Grey Ghost* due to its flat-gray paint job) again in front of a large crowd of workers and VIPs. Everyone had a good time except LeVier, who wound up having a pretty exciting day.

Upon takeoff, he discovered that the pressurization system was pouring 100-plus-degree air into the cockpit. Once airborne, he discovered the aircraft was unstable and almost uncontrollable. It took several minutes before he could attain a speed of 260 mph to start his climb. On reaching 10,000 feet, he began the test card items, one of which required him to extend the flaps. The aircraft quickly rolled inverted due to the fact that only one flap came down. He decided that was enough for one day and returned to land after only 20 minutes, holding full right stick to remain upright during the approach.

As he stated in the flight report, "Thank goodness for the hydraulic aileron boost!"

That instability was caused by the fuel load being in error and the airplane's center of gravity being too far aft. *Grey Ghost* turned out to be somewhat of a "hard luck" airplane and did not last very long. On March 20, LeVier was flying a high-speed test when turbine blade failure caused the airplane to lose its entire vertical tail section after completing a high-speed dive to 560 mph. The jet tumbled violently, and without an ejection seat, LeVier assumed he would not survive. Suddenly, the aircraft stabilized momentarily, and he was able to jettison the canopy and bail out over the side.

LeVier's exit was tumultuous, as was his landing, and he broke his back during his parachute landing fall. There had been a metallurgical fault in the turbine wheel, and when it failed it acted like a buzz saw, neatly cutting off the entire tail section. *Grey Ghost* was

Lockheed XP-80 is parked on Rogers Dry Lake. The color scheme was a semi-gloss dark Kelly Green with gray undersides as a result of the secret jet aircraft passing through Lockheed's paint facility in Burbank at the same time twin-engine Hudson patrol bombers were being painted for the Royal Air Force. (Wings & Airpower Historical Archive)

U.S. Air Force maintenance and General Electric tech support personnel mount this 5,400-pound-thrust General Electric I-40 centrifugal-flow turbojet on a rolling work stand after its removal from a Lockheed P-80 Shooting Star. (Wings & Airpower Historical Archive)

Lockheed test pilot Tony LeVier flies the sleek XP-80A Gray Ghost over Rogers Dry Lake, California. Lockheed P-80R flown by Col. Al Boyd was specially modified with a more powerful jet engine, a smaller, low-drag canopy, and water-methanol injection to exceed the British records by the required 1 percent, which the standard P-80 was not capable of accomplishing. The average speed over his four runs on 19 June 1947 was 624 mph, the first U.S. record over 1,000 km/h and America's first official world speed record in 25 years. (Wings & Airpower Historical Archive)

Lockheed's XP-80A is in a most unusual chase plane view flying over the alfalfa fields of the Antelope Valley near Muroc Flight Test Center. This view gives a graphic depiction of the aircraft's sleek design, stemming from the genius of Kelley Johnson and his revered Skunk Works advanced design group. (Wings & Airpower Historical Archive)

gone, but had served its purpose in discovering a number of faults in the aircraft's design at high speeds. One of these was an aileron buzz that developed at 0.8 Mach. The hydraulically boosted ailerons were still controlled by cables, and engineers were concerned that at high speed, aileron buzz might increase in magnitude until the ailerons came off. It was discovered that by simply tightening the cables, the buzz disappeared.

The second experimental prototype XP-80A flew in July 1945 equipped with a second seat for a flight test engineer, which replaced a fuel tank. Pilots reported hearing what sounded like a rumble coming from the air inlet ducts and Kelly Johnson squeezed into the small second seat to listen for it in flight himself. He did indeed hear it and decided the rumble was probably caused by turbulent airflow from the boundary layer in the ducts. He modified the inlets with vented "gills" to siphon off the boundary layer, and as a result, the duct rumble ceased.

While XP-80As were undergoing advanced tests, the newly formed U.S. Air Force ordered pre-production prototype YP-80As for initial service tests. Although these aircraft were similar to the prototype XP-80A, they more closely resembled the final configuration of the operational P-80 jet fighter. Unfortunately, the YP-80As suffered casualties due to flameouts, usually right after takeoff from Burbank Airport. The problem was eventually traced to an unreliable fuel pump and later aircraft were modified with a secondary back-up electrical fuel pump.

Sadly, famed Lockheed test pilot Milo Burcham was killed on his first flight in one of these aircraft, and later Maj. Richard Bong, America's highly celebrated top-ranking World War II ace, died in the crash of his YP-80A on Lankershim Boulevard south of Burbank Airport after the back-up pump also failed. Despite these setbacks, consideration was given to deploying P-80s to Italy, but the war ended before that plan could be carried out. Instead the P-80 saw combat in a different part of the world in 1950 when the Korean War broke out.

The Skunk Works' participation in developing P-80 prototypes was coming to an end, but there was one more task to be accomplished as the war ended and world speed records again could be officially established. The British Meteor raised the bar by setting speed records that for the first time exceeded 600 mph. Lockheed and the Air Force decided it would be possible to break the British hold on those records with the P-80, but it would be necessary to modify a standard Shooting Star to break the record with sufficient margin to establish a new world record. With the war over, the race was on to set new records; now they faced not only other competitors, but also the mysteries of something call the "Sound Barrier."

Representing the absolute zenith of piston-powered aerodynamic streamlining was Republic's XR-12 Rainbow. Designed at the end of World War II as a dedicated high-speed, high-altitude photo-recon platform, the XR-12 was the world's fastest piston-powered four-engine aircraft with a top speed of 462 mph at 40,000 feet. Despite this impressive performance, the airplane was still too slow and too vulnerable to attack by jets, but too fast (and thus, too expensive) to be a successful airliner. (Republic Aviation Corporation)

A beautiful example of early jet aircraft design is this Hawker P.1040, a straight-wing twin-jet with cruciform tail, and forerunner to the famed Royal Navy Sea Hawk. Just as Republic's XR-12 represented the end of piston-powered supremacy in aircraft design, airplanes such as the P.1040 represented the very beginning of practical and efficient jet aircraft technology. Its straight wing, however, was soon obsolete. (The Museum of Flying)

Chapter 4

BREAKING THE SOUND BARRIER (1946–1956)

Nearly every possible aerodynamic configuration is represented here with the "X-Planes," the first generation of exotic U.S. experimental research aircraft that flew from 1947 to 1956. Clockwise from lower left are: Bell X-1A, first aircraft to fly faster than Mach 2; transonic Douglas D-558-1 Skystreak; delta-wing Convair XF-92; swing-wing Bell X-5; swept-wing Mach 2 Douglas D-558-1 Skyrocket; tailless Northrop X-4 Bantam; and needle-nose Douglas X-3 Stiletto, center. (USAF)

Where it all happened in the heady post-war years, the USAAF/U.S. Air Force Flight Test Center at Muroc, California, which became Edwards Air Force Base in 1949. Shown here is the South Base complex, site of all the great rocket-plane flights and development center for every new USAF aircraft of that era. In 1955, the large area at the upper right became the new "Main Base" or "Super Base" with a 15,000-foot runway and many miles of facilities. It is the current home to the Air Force Test Center and NASA Armstrong Flight Test Center. (Mike Machat Collection)

The opening narration of the iconic aviation movie *The Right Stuff* begins with the words, "There was a demon that lived in the air." Spoken eloquently by actor and musician Levon Helm, who plays Air Force Capt. Jack Ridley, sidekick to the legendary Chuck Yeager in the film, these words denote the supposed pilot-killing "barrier" that stood between the speed of sound and mortal man. Identified by the even more cryptic name "Mach 1," this speed seemed all but unattainable in the years following World War II, as the assault to conquer that demon began in earnest on both sides of the Atlantic.

The Quest for World Speed Records in the Jet Age

The Skunk Works participation in development of the Lockheed P-80 prototypes was coming to an end, but there was one more task to be accomplished after the war as world speed records could again be established officially. The British struck first on 7 November 1945 with a Gloster Meteor F Mk. 4 flown by Capt. H. J. Wilson, setting a record for the first time over 1,000 kmph, or 606 mph. As the absolute speed record had to be set below 100 m (330 feet) altitude, the Royal Air Force elected to make the flights over the ocean at Herne Bay. At low altitudes, the cool, dense air increased the mass flow and thrust of the two jet engines, which helped set the record. The Meteor beat the pre-war Messerschmitt 209's old record by almost 30 percent. Aircraft used for the record flights were further modified with long-chord nacelles and longer wings to increase their speed margin over the standard Meteor F. On 7 September 1946, the record was increased to 615 mph.

Lockheed and the Air Force decided it would be possible to capture the speed record with the P-80. It was necessary to modify a standard P-80 to achieve a required speed one percent higher than the existing record. The had British raised the bar, so the new goal

This inflight photo illustrates the 18-degree sweep back on the wing of the Messerschmitt Me-262. This had nothing to do with sweep back to delay critical Mach number, but rather was necessary to reposition the aircraft's center of gravity when the design was switched to accommodate twin jet engines. (National Archives courtesy of Dennis R. Jenkins)

The zenith of multi-engine piston-powered fighters was Lockheed's P-38 Lightning, used in World War II when aircraft flew fast enough to experience compressibility problems in steep high-speed dives made from high altitude. P-38s experienced control difficulties in these dives as the airplane reached near-sonic speed and airflow over its wings and tail began to produce shockwaves. It took aircraft powered by rockets and then more powerful jet engines coupled with swept wings and tails to solve this problem, allowing them to reach supersonic speed. (Wings & Airpower Historical Archive)

was 621 mph. This was beyond the capability of the standard P-80 so Lockheed extensively modified the remaining airplane. It received the designation P-80R (for Race). It had a smaller canopy to minimize the drag as well as a new, more powerful J33 turbojet with 4,600 pounds of thrust. A major modification was installation of water-methanol injection for the turbojet engine.

On 19 June 1947 all was ready and Col. Alfred Boyd, air force chief test pilot at Muroc, flew over Rogers Dry Lake on the 3-km course, generally staying 200 feet above the ground and relying on red smoke markers for navigation purposes. His speeds were measured officially at 617, 614, 632, and 631 mph, giving an average speed of 624 mph. America now had its first official world speed record in almost 25 years.

Approaching the Sound Barrier

The high speeds made possible with jet engines proved to be a two-edged sword as far as setting new records was concerned. The

NACA's" flying wind tunnel" used the top of the P-51 wing surface while in a steep dive to create transonic airflow passing over the small wind tunnel model mounted atop the wing. At these speeds an actual ground-based wind tunnel would have shockwaves forming and bouncing off the walls, thus providing inaccurate data. In contrast, the free-flying P-51 was not affected by reflected shockwaves. (NASA)

NACA's Northrop RF-61C carries a streamlined shape to be dropped from high altitude to gather transonic data on aerodynamic pressures. Data was sent to a ground station via radio telemetry during the 60-second descent from 40,000 feet. The shape was low drag, and therefore went supersonic at some point before hitting the ground. (NASA)

First-generation jet fighters incorporated turbojet engines housed in a long cylindrical fuselage, but with straight wings and tails evolved from World War II aircraft design. Engineers designing the Republic F-84 Thunderjet were quoted as saying, "We were actually writing the book on jet aircraft design as we built these airplanes." The F-84 was a subsonic aircraft with a top speed of 620 mph, and nearly 5,000 of them were built. (Republic Aviation Corporation)

First evolution of the Thunderjet was the YF-96 (later F-84F) Thunderstreak, which adapted swept wings and tailplanes to the basic F-84 fuselage. This configuration allowed the jet to become a transonic airplane with a top speed of 710 mph, and later versions with "all-flying tails" were capable of barely achieving supersonic speed in a dive. Both the Thunderjet and Thunderstreak were flown by the Air Force Thunderbirds Aerial Demonstration team. (Republic Aviation Corporation)

This 20-foot mural at the Air Force Flight Test Center Museum at Edwards AFB illustrates the Golden Age of Flight Test. Leading the formation are the legendary rocket planes, followed by their carrier aircraft with appropriate chase airplanes. Oddly shaped aircraft that pioneered different wing shapes for the Jet Age are shown in the bottom row. Diving for the lakebed below the formation is an X-24 lifting body, followed by its NASA F-104 chase. The vista at the bottom shows Edwards in the glory days of the early 1960s. (Mike Machat)

The first-generation Bell X-1 was originally designated as the "XS-1" (for Experimental Supersonic), and was designed exclusively to attack the sonic barrier. Shown here is Ship 2 (46-063), the airplane flown by Bell's colorful test pilot Chalmers "Slick" Goodlin. This aircraft had a 10-percent wing thickness ratio compared to the air force's 46-062 flown by Capt. Chuck Yeager, which had an 8-percent thickness with commensurately higher stall and landing speeds.

ability to fly level at speeds of more than 600 mph meant that the aircraft itself was approaching the speed of sound. A function known as compressibility demonstrated during World War II at high altitudes was now coming into play for low-level high-speed flights, and the phrase "breaking the sound barrier" became a popular phrase known vaguely to the general public.

The "sound barrier" was the speed of sound in air, which is actually a function of temperature. This speed is approximately 760 mph at sea level but decreases to 660 mph in the cold, thin air of the stratosphere. With the advent of powerful high-speed jet aircraft, designers and, indeed, pilots now had to be concerned about the speed of the airplane compared to the speed of sound at high altitudes. This percentage of the speed of sound received the designation "Mach" number, named for the Austrian physicist Ernst Mach, who first explored the effects of temperature on the speed of sound.

As an airplane approaches the speed of sound, strange things begin to happen to the airflow around that airplane, which drastically affect both drag and controllability if the airplane is not designed to operate at those high speeds. The investigation of aircraft behavior and proper design in this unexplored region was a major technological challenge that in turn had a major effect on aircraft design and spurred new interest in the race to set world speed records. Since jet engines were initially extremely expensive (and still are), the competition to set new speed records switched from private citizens to a competition between countries and their respective military air forces.

In the aftermath of World War II, only the British and the United States had jet engines with which to set the records, although as the Cold War developed, the Soviet Union also acquired jet engine technology. Great Britain set the first jet speed records, but American engine technology soon matched that of the British and speed records were set mainly by Americans. However, an exploration of supersonic flight had to occur before new higher speed records could be set. New engines and even new aircraft shapes had to be developed. The tale of that investigation and subsequent results shaped aviation for the remainder of the 20th Century.

High-Speed Flight Test Research

Prior to the 1930s there was no concern about Mach number or compressible flow because aircraft of the era were incapable of flying at velocities greater than about Mach 0.5. Nevertheless, aerodynamicists realized there might be a problem for high-speed aircraft in the future and held a conference in Rome in 1935 where theoreticians discussed what would happen as airplanes moved to higher Mach numbers. In approaching the speed of sound, studies suggested, a low thickness-to-chord-ratio wing (less curvature of the top surface) allowed the airplane to fly closer to the Mach 1 before airflow reached the speed of sound. This critical airspeed became known as the "critical Mac number."

The fairly austere instrument panel of the Bell X-1 included basic flight instruments and a Machmeter that only went to Mach 1.0 (upper right). Pilot entry and egress was through a small hatch to the right of the cockpit, which pretty well ensured there was no safe way to manually bail out of the airplane. (Wings & Airpower Historical Archive)

The critical Mach number was also known as the "drag divergence" Mach number because the formation of shockwaves attached to the wings caused a steep increase in overall drag. German aerodynamicist Adolph Busemann further suggested that by sweeping the wing by an appreciable angle, the airflow over the wing would actually be divided into chord-wise flow (from leading edge to trailing edge), which generated lift, and span-wise flow parallel to the

The Speed of Sound in the Atmosphere

The speed of sound in the atmosphere is mainly a function of temperature. The table lists representative values for a "standard day" (empirically based on a specified sea level temperature and growing cooler by two degrees centigrade to the tropopause [approximately 36,000 feet altitude] where the temperature then stabilizes for tens of thousands of feet before becoming much less predictable at even higher altitudes.

For jet- and rocket-propelled aircraft, speeds are often given as Mach numbers or percentages of the speed of sound, with a Mach number of "1.0" being the local speed of sound.

Speed of Sound at Different Altitudes					
Altitude	Temperature	m/s	kmph	mph	kn
Sea level	15°C (59°F)	340	1,225	761	661
33,000 to 66,000 feet (Cruising altitude of commercial jets, and first supersonic flight)	−57°C (−70°F)	295	1,062	660	573
90,000 feet (Ceiling of SR-71)	−48°C (−53°F)	301	1,083	673	585

Compressibility, Shockwaves and Supersonic Flow

The Earth's atmosphere consists of a gas mixture of nitrogen and oxygen, with certain pressure, density, and temperature. Atmospheric motion tends to remain relatively low-speed with winds of no more than 200 mph, and occurring only at extremely high altitudes where the air is low temperature at low density. At sea level, winds seldom exceed 40 mph, except during the occasional violent storm. Airplane lift is generated by moving through the atmosphere at higher velocities, thus reducing the pressure of the relative wind airflow over the top of the airplane wing. At low relative wind velocities, the density of the air remains constant, and this is known as "incompressible flow."

However, as an aircraft moving through the atmosphere increases its velocity, the density of the air around that aircraft is actually lowered, as is the pressure of the relative wind approaching the speed of sound. This change in density is known as "compressible flow," and when the relative wind around the aircraft reaches and exceeds the speed of sound, shockwaves occur around the body. In shockwaves the pressure and density of the airflow change drastically to slow that flow down to the speed of sound or below. For airplanes, the speed of sound is a natural speed limit, and shockwaves form near the nose and the tail as the atmosphere attempts to return to velocities less than the speed of sound.

Airplane wings are shaped to generate low pressure over the top of the wing for lift, which is accomplished by causing the airflow on the top of the wing to move faster than that on the bottom. Therefore, as the airplane approaches the speed of sound, the top of the wing achieves a speed of Mach 1, although the airplane itself is actually flying at less than the speed of sound. Once the entire airplane is flying faster than the speed of sound the airplane is in a supersonic flow region where compressible flow theory can more accurately predict the pressures that generate lift and drag.

Before reaching supersonic flight there is a transonic region, defined as from Mach 0.8 to Mach 1.2, where the airflow is extremely complex, as it is a mixture of incompressible and compressible flow, which is much more dynamic and difficult to analyze. The "sound barrier" actually refers to the transonic speed region where shockwaves first form on the airplane and where there is a steep increase in drag. Flying supersonic requires flying through the transonic speed regime first, which initially proved to be difficult due to the chaotic dynamic flow combined with the fact that wind tunnels originally could not provide reliable data. This was due to shockwaves forming in the test section, which then reflected off the walls of the tunnel, providing incorrect pressure data.

Shockwaves in air are quite similar to ripples emanating from the bow of a boat moving through water. The faster a boat goes, the bigger the ripples. As an airplane moves through the air it sends out signals to the atmosphere air molecules to move out of the way. Those signals move through the atmosphere at the speed of sound. As the airplane approaches the speed of sound the plane arrives before the molecules receiving those signals move out of the way. This causes air molecules to "pile up" in front of the airplane in a very thin region (less than an inch thick). In this region, the density of the air changes as well as the pressure of the atmosphere.

These regions of shockwaves exist in two types: normal shock and oblique shock. A normal shock exists at 90 degrees to the flight path of the airplane's airstream, and on the downstream side of the shock the airflow is subsonic. On the upstream side of the shock the flow is supersonic, or faster than Mach 1. The oblique

Interesting study of shockwave propagation as a U.S. Marine Corps McDonnell Douglas F-4S Phantom II goes supersonic in the vertical while conducting aerial combat maneuvering in humid conditions over the Atlantic Ocean. Shockwaves were observed completely encircling the fuselage (shown) as well as on the leading edges of the air intakes. (Mike Machat)

shock is at some angle less than 90 degrees to the airflow, and on both the upstream and downstream side of the shock the airflow is still supersonic.

An airplane flying in the transonic region generally has a normal shock attached to the wing. The pressure change through the normal shock causes the airflow aft of the shock to separate from smooth laminar flow into turbulent flow, creating much drag and fluctuations in pressure, which also lead to loss of lift. This loss of airflow smoothness is known as a shock stall, which is one reason why it is so difficult to fly at the speed of sound due to the increased drag in the transonic regime from the strong influence of a normal shock.

For an airplane flying at Mach 2, a properly designed wing has oblique shocks predominate on the wing, rather than a normal shock, and drag is less than if flying at Mach 1. Thus the sound barrier is actually more of a "drag barrier." The Mach-3 SR-71 can accelerate through the speed of sound in level flight, but operational pilots discovered that by diving at a shallow angle approaching Mach 1 to penetrate the transonic regime faster, less fuel was used, which translated to extended range at Mach 3.

The shock stall in the transonic speed region reduced lift not only on the wing, but also on the elevator for the normal horizontal tail. This fluctuating lift at Mach 1 caused buffeting, which made the aircraft difficult to control. Hydraulics connected to the control surfaces of high-speed airplanes were beginning to be con-

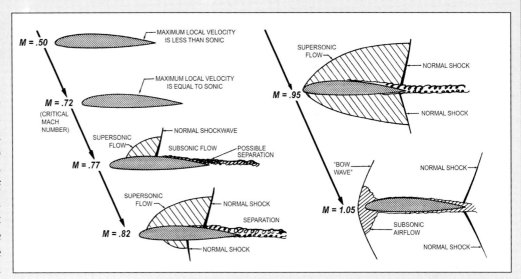

This chart shows subsonic and supersonic flow properties. (Graphic technical enhancement by Craig Kaston)

sidered to aid the pilot in physically moving the control surfaces at high speeds. This also became an accepted practice in jet aircraft to damp out control surface buffeting near Mach 1.

An even more accepted practice was to eliminate the elevator on the horizontal tail surface, instead moving the entire horizontal tail for pitch control. Test pilot Chuck Yeager often mentions this as being the most important discovery from the Bell X-1 program because the "all-moving" horizontal tail was installed on later models of the North American F-86 Sabre, which greatly increased its maneuverability and combat effectiveness against the Soviet MiG-15 in Korea. By comparison, the MiG retained elevators on its horizontal tail.

The modified Boeing B-29 was designated as the EB-29 mothership (later named "Fertile Myrtle") that carried the first-generation Bell X-1s to launch altitude. Although difficult to discern in this photograph, both pairs of main wheel landing gear doors were removed to provide clearance for the X-1's wingtips. The black square at the nose of the X-1 is the manually operated elevator for pilot entry in flight. (Wings & Airpower Historical Archive)

Here, 24-year-old Air Force test pilot Capt. Charles E. "Chuck" Yeager poses in the late afternoon of 14 October 1947 after his first flight to Mach 1. The photographer's shadow on the X-1 fuselage is that of Yeager's best friend and fellow air force pilot Capt. Jack Ridley. (Courtesy of the AFFTC History Office)

leading edge of the wing, which did *not* generate lift. The high-speed wing would then be "fooled" into having chord-wise flow along the top of the wing moving at less than the speed of sound, even though the aircraft itself was actually flying faster than the speed of sound, thus avoiding the formation of shockwaves on the wing.

For example, at Mach 1, the chord-wise flow of a 35-degree wing would be Mach 0.82, while respective figures for a 45- and 60-degree swept-back wing would be Mach 0.7 and Mach 0.5. This suggestion was duly noted in the technical papers from the conference, but as of 1935 no airplane was considered capable of reaching the speed of sound. The suggestion was filed away and promptly forgotten, even by the international group of famed aerodynamicists who actually attended the conference.

During World War II, fighters flew fast enough to experience compressibility problems in steep high-speed dives made from high-altitude. The Lockheed P-38 Lightning and Republic P-47 Thunderbolt both experienced difficulties in high-speed dives at high altitudes. P-38 pilots reported, "It felt like the stick was set in concrete and the airplane pitched over into a steeper and steeper dive, and I couldn't pull out of it until I got to a lower altitude." No one understood the reason for this behavior, but Lockheed developed a unique "drag flap" that operated via an electrical switch to allow the pilot to regain control of the airplane.

The drag flap, located in mid-chord on the underside of the wing, did not produce as much drag as did traditional dive bomber dive brakes. For most NACA wing profiles, the lift force center of pressure shifted aft from the quarter-chord position to approximately the half-chord position once shockwaves formed on the wing. This phenomenon occurred both on the P-38's large wing and wide horizontal stabilizer with its movable elevator to control pitch.

This new arrangement of supersonic lift generated on both the wing and tail caused the airplane to pitch over into an even steeper dive with the shockwaves on the tail rendering the elevator at the trailing edge even less effective, preventing the pilot from pulling out of the dive. As the P-38 descended to lower altitudes and higher temperatures, the speed of sound increased and the airflow over the

The Bell X-1 breaking the sound barrier became front-page news on 22 December 1947 when it was finally revealed that man had flown faster than the speed of sound. This highly guarded information had been classified previously. This is the Los Angeles Times depiction of the secret "U.S. Mystery Plane" that smashed the sonic wall and propelled aviation into the supersonic era. (Wings & Airpower Historical Archive)

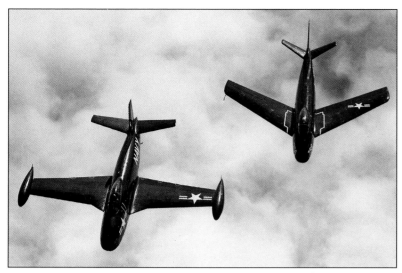

A good depiction of North American's jet fighter designs post–World War II. The FJ-1 Fury (left) was built with a straight wing and was somewhat underpowered. It was not successful other than as a jet familiarization tool. The FJ-2 Fury, on the other hand, was based on the swept-wing F-86 Sabre and versions remained in service well into the late 1950s. Swept-wing jet fighters for the air force and the navy were worthy successors to North American's landmark P-51 Mustang. (U.S. Navy)

P-38's wing and control surfaces was no longer supersonic. The center of lift shifted back to the quarter-chord position, and lift distribution was restored to normal, allowing the pilot to regain control of his aircraft. The innovative small drag flap when deployed created just enough drag to disrupt airflow over the wing to suppress shockwave production while still at a higher altitude. It worked well, even though no one at the time understood exactly why.

The Swept Wing and Transonic Flight

Translations of captured German research papers after the war indicated that sweeping the wings of an airplane was a way to delay the onset of the drag from shockwaves attached to the wing during transonic flight. A by-product of this design meant that takeoff and landing speeds and distances would be higher unless high-lift devices such as slats on the leading edge and flaps on the trailing edge were installed on the wing. In addition, the presence of span-wise flow on a swept wing caused disturbances near the wing tip, resulting in a tip stall where the ailerons were located, reducing their effectiveness for banking the airplane. In addition, at high speeds the stalled outer portions of the wings caused a pitch-up as the wing's lift distribution changed. A swept-wing testbed aircraft was needed to investigate the effects of wing sweep on an airplane's flight characteristics.

Observers also noticed that the Messerschmitt Me-262 and Me-163 both had swept wings. The reason for this was not for high speed. The center of gravity of the Me-262 had to be shifted aft once a second heavy jet engine was installed, and the rocket-powered Me-163 was a tailless flying wing where the ailerons doubled as elevators, they had to be shifted aft via wing sweep back to function as elevators. The Germans actually did not build a swept-wing airplane for critical Mach reasons, but the Messerschmitt P 1011 jet prototype (which was never completed) did indeed have a swept wing for increasing critical Mach number.

Shockwave action on the models tested in the original ground-based supersonic wind tunnels often resulted in incorrect data. It seemed the only way to gather real data in the transonic speed regime was to accomplish some sort of flight test. Building a full-scale test airplane required creative thinking, and calculations for an experimental aircraft dictated a need for thrust that was not yet achievable with the anemic turbojet engines of the mid-1940s.

This in turn dictated the use of a rocket engine with its high thrust but extremely short engine burn time, as demonstrated by the Heinkel He-176 and Me-163 during the war. This short flight duration meant that the airplane would have to be flown by an extremely efficient (and extremely dedicated) test pilot. This eventually led to the golden era of the X-planes and new test pilot heroes to replace the much-admired race plane pilots of the 1920s and 1930s.

Wing Research and Design in the Transonic Era

Investigating the transonic speed regime in flight required innovative solutions. One approach was to install small model wings on the top surface of high-speed fighter plane wings, which would then dive from high altitudes at Mach numbers of 0.9 or higher. The higher-velocity air over the top of the wing approached the speed of sound for very brief periods of time, and the test aircraft would be instrumented with pressure taps to collect data on velocities and pressures as shockwaves formed around the models.

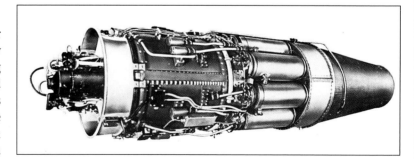

This engine represented the dawning of a new age in turbojet design, the axial-flow General Electric TG-180, designated as J35. Capable of producing 3,750 pounds of thrust, it ushered in a new era of high-performance flight, powering such aircraft as the Republic F-84 Thunderjet, Vought F7U Cutlass, and Douglas X-3 Stiletto. The engine's smaller diameter enhanced aerodynamic streamlining. (Wings & Airpower Historical Archive)

It's easy to see why this airplane was nicknamed "The Crimson Test Tube," as Douglas test pilot Gene May poses with the record-breaking D-558-1 Skystreak on the Lakebed at Muroc. Because the Skystreak and its successor, the D-558-2 Skyrocket, were initially navy programs, they did not receive "X" designations as did their air force counterparts, but they were in every sense part of the hallowed family of Edwards' X-Planes. (Wings & Airpower Historical Archive)

Later-configuration cockpit on the Skystreak in NACA colors. Designed to probe the nether regions of transonic flight, the D-558-1 with its straight-wing, perfectly cylindrical fuselage, and uprated 5,000-pound-thrust J35 turbojet was never intended to fly at supersonic speed, but provided valuable research data in the quest for reaching Mach 1.

Another somewhat less hazardous test method was to drop streamlined shapes from high-flying larger aircraft with a modicum of data being transmitted to the ground via radio. A streamlined bomb-shape dropped from high altitudes will generally reach supersonic speed by the time it impacts the ground, as the time of free-fall from 40,000 feet is approximately 60 seconds.

A third approach to transonic flight investigation was to mount instrumented models atop solid fuel rockets and fire them into the stratosphere from the NACA test facility at Wallops Island on the Chesapeake Bay, close to the Langley aeronautical labs in Virginia. This "flying wind tunnel" approach provided longer data collection times, but was very expensive because the models for each experiment crashed into the sea. To attain a more economical and repeatable way of collecting data, a full-scale research airplane capable of transonic speed in level flight had to be built.

Piloted Transonic Research Aircraft: The Early Days

Transonic flight using new jet engines with the wing designs developed in the early 1940s indicated two problem areas. More thrust was needed than was available in current jet engines, and there was a serious controllability problem on prewar aircraft when approaching the speed of sound. To operate an experimental aircraft at transonic speed long enough to collect flight test data, researchers decided that the aircraft should be taken to altitude by another airplane and released in flight, allowing the rocket thrust to be used only to reach speeds where meaningful data could be collected.

The need to be carried by another airplane was somewhat of a limitation for the initial experimental airplanes, such as the Bell X-1 and the Bell X-2, affecting their overall size. As a result, they had small fuel tanks with rocket burn times on the order of between 1 and 2 minutes. A similar British project known as the Miles M 52 was proposed as early as 1943 to investigate transonic flight, and the aircraft's design was mainly completed when it was canceled in 1946.

The British were initially suspicious of liquid-fuel rocket engines and proposed using a high-thrust jet power plant, meaning the aircraft could take off from the ground under its own power. The jet engine planned for use in the Miles M 52 was innovative in itself, being a turbofan engine with an afterburner, two features that in later years proved to be pivotal in increasing engine thrust. A test pilot had actually been chosen to fly the airplane, Capt. Eric Brown, a naval officer who had test flown numerous aircraft, including the German jets. He also flew the Me-163, but usually as a towed glider, because use of the rocket motor was considered too dangerous.

The Miles M 52 design was reminiscent of the Bell X-1, with straight wings having a low thickness-to-chord ratio. The aircraft had a tail, but rather than having an elevator attached to a horizontal

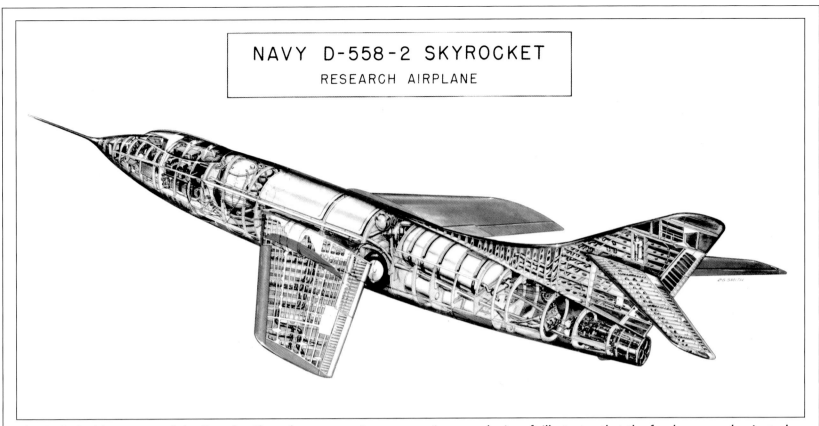

NAVY D-558-2 SKYROCKET
RESEARCH AIRPLANE

This R. G. Smith cutaway of the Douglas Skyrocket swept-wing supersonic research aircraft illustrates that the fuselage was dominated by fuel and oxidizer tanks for the rocket engine. The turbojet relied on the oxygen in the air surrounding the airplane while the rocket engine had to bring its own air with it, thus explaining the limited amount of time the engine ran. (Douglas via National Archives)

The Douglas D-558-2 Skyrocket is the only aircraft ever built to fly as a straight turbojet, pure rocket, and hybrid turbojet/rocket-powered airplane using the same airframe. Shown here making a JATO takeoff from Rogers Dry Lake, the Skyrocket was initially flown as a pure jet, although its single Westinghouse J34 power-plant provided only a modicum of the thrust needed for true high-speed flight. (Wings & Airpower Historical Archive)

P2B-1S mothership, the navy version of Boeing's venerable B-29, launches the Douglas D-558-2 Skyrocket. The bomber's World War II–vintage color scheme of black undersides was used in wartime to make it difficult to see from the ground at night if searchlights briefly illuminated the airplane, but these markings were retained for its flight test work. (Douglas Aircraft Company via Mike Machat)

Marine test pilot Col. Marion Carl poses with the Skystreak, very possibly for a Wheaties cereal box. Note the cloth World War II–type helmet, still in use in the late 1940s. The Skystreak's limited canopy size caused tall pilots to prefer cloth helmets to avoid constantly bumping the canopy with larger hard helmets when flying in turbulence. Carl set several speed and time-to-climb records while at Muroc. (U.S. Navy)

stabilizer, the entire horizontal surface rotated to control pitch. British and American cooperation in high-speed flight developed during the war meant that when the Miles M 52 was canceled, the data on the design of the airplane was shared with the Americans, including the use of the "all-flying" horizontal stabilizer to control pitch.

The Bell X-1's fuselage design was based upon a shape that was known to be able to go supersonic: the Browning .50 caliber machine gun projectile with a muzzle velocity of approximately Mach 3. Based upon previous aircraft design studies, two sets of wings were planned for the X-1 with different thickness-to-chord (T/C) ratios for the wing of 10 and 8 percent, and 8 and 6 percent for the corresponding horizontal stabilizers. A movable horizontal stabilizer was included, but it was intended to be used only to trim the vehicle, so a conventional movable elevator was installed, connected to the control wheel.

When the X-1 approached transonic speed for the first time at an indicated Mach number of 0.94 the elevator became ineffective. The test team concluded that it might use the movable horizontal stabilizer to continue increasing the speed to collect data, and on 14 October 1947, the rocket-powered Bell X-1 exceeded the speed of sound in level flight for the first time. Progress was being made toward collecting data that promoted safe flight through the sound barrier.

The X-1 was first planned in 1943 with straight wings, because German research into the benefits of wing sweep in the transonic speed regime did not become available until the spring of 1945. Engineers at NACA and in the American aviation industry read the theoretical German papers that had been translated and realized that thin swept wings would allow operational aircraft to operate in that speed regime using jet engines that were available or soon would be available.

By sweeping the wing at 30 to 35 degrees it was theoretically possible to fly at high subsonic and possibly low supersonic speeds without the formation of supersonic shockwaves greatly increasing the drag on the wing. Airplanes built in this immediate postwar era were generally not capable of supersonic speed in level flight, but during combat operations in the stratosphere, fairly shallow dives could result in airplanes approaching the speed of sound.

Navy and NACA Cooperation Conquers Mach 2

While the U.S. Army Air Forces arranged to cooperate with NACA on the construction of the Bell X-1 in 1943, the navy also wanted to investigate transonic speeds in cooperation with NACA. In 1947, the

As with so many successful aircraft designs, the airplane in its original form often little resembles the operational configuration. This is the D-558-2 Skyrocket at Douglas's Santa Monica plant with its original flush canopy. Although aerodynamically perfect as far as streamlining for high speeds, this design was severely limiting for pilot visibility. A canopy with a raised V-shaped windshield was the answer. (Douglas Aircraft Company via Mike Machat)

Famed Douglas Chief Engineer Ed Heinemann examining Skyrocket's leading-edge slats. Adjacent to him is a wing fence installed to prevent span-wise flow from causing wingtip stall prematurely. Various configurations were tried during the Skyrocket's flight test career. Note the flush-mounted canopy initially proposed for the Skyrocket but quickly replaced due to poor visibility for the pilot, especially at high angles of attack. (Douglas via National Archives)

Douglas Aircraft Company was chosen as the builder for the D-558 project, as it was known at Douglas. The air force was in charge of handing out "X" designations, so the navy project retained the original internal Douglas designation. In order to gather more transonic data, a jet engine would be used so the airplane could fly for up to an hour, versus only the several minutes of the rocket-powered flight with the X-1.

Initially six D-558 jets were to be built to investigate flight speeds below Mach 1. Phase 1 would explore transonic speeds, with Phase 2 using a more powerful engine to exceed the speed of sound in level flight. Based upon the German research, the last three airplanes were switched to swept wings and speeds were now to go beyond Mach 1. The Phase 1 airplanes, called Skystreaks, had straight wings and turbojets while Phase 2 airplanes, called Skyrockets, had 35-degree swept wings. The third Skyrocket was originally equipped with a jet engine for takeoff, but was later converted into a pure rocket airplane to be air-launched from a navy Boeing B-29.

The D-558-1 Skystreak was painted bright red and became known as "the crimson test tube" because of its flying laboratory mission and cylindrical fuselage that literally looked like a test tube with a cockpit, straight wings, and tail attached to it. The joint navy/NACA program collected much technical data and the three Skystreaks flew for almost 300 flights, providing excellent data as testbeds investigating high subsonic flight, but also such advanced technical innovation as winglets. These were to investigate methods of producing wings with higher lift/drag ratios than high-speed low-aspect-ratio

Looking like a record breaker from any angle, this stunning view of the sleek Douglas Skyrocket shows why it was nicknamed "Flying Swordfish." The D-558-2 was the first airplane to fly Mach 2 and above 80,000 feet, and has the distinction of setting records piloted by civilian contractor, U.S. Marine Corps, and NACA pilots. (Douglas Aircraft Company via Mike Machat)

wings usually produced. Winglet research also revealed that installing them restricted span-wise flow, a fact that became significant on swept-wing airplanes.

Because they could take off under their own power and had sufficient endurance, Skystreaks were deemed as suitable in making the required four speed runs to set official FAI speed records. In August 1947, Skystreaks were flown by navy and Marine Corps' pilots trying to break the army air force's P-80R speed record. Rather than having to be extensively modified as the Lockheed airplane was, the Skystreak, using a higher-thrust J35 axial-flow engine with 5,000 pounds thrust, was able to reach speeds of 641 and 651 mph on the Muroc course while carrying 640 pounds of special instrumentation.

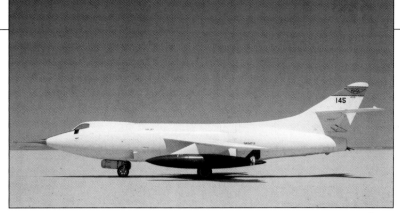

Although we almost take it for granted that supersonic aircraft carried external auxiliary fuel tanks, these devices had to be flight tested before becoming operational on military fighter aircraft. It was the rugged and dependable D-558-2 Skyrocket that served as the critical testbed for generating research data leading to the certification of supersonic fuel tanks. (Wings & Airpower Historical Archive)

Marine test pilot Col. Marion Carl also used the Skystreak to set time-to-climb records during his stay at Muroc.

The D-558-2 Skyrocket was the first airplane to reach Mach 2 and is displayed today in the National Air and Space Museum because of that achievement. Its real value was not just in the records it set, but in the pioneering work it did exploring the flight characteristics of a thin swept wing. Its wings were modified several times with installations of wing fences and modified outboard leading edges to investigate preventing span-wise flow from causing undesirable flying characteristics such as an un-commanded pitch-up. These wing fences were one way of blocking span-wise flow before it reached the wing tip and stalled-out the ailerons.

Many swept-wing jet fighters, especially in the Soviet Air Force, used wing fences, with the MiG-19's radical 59-degree swept wing having monstrously high fences. (The MiG-19 was Russia's first jet fighter to exceed Mach 1 in level flight.) By comparison, the Douglas D-558-2 Skyrocket with its slim tapered fuselage, needle nose, and swept wing represented exactly what a jet plane should look like in the fabulous 1950s. It was nicknamed the "Flying Swordfish" and even gained fame as Capt. Midnight's new ride when the popular World War II cartoon and radio show moved to television.

Not to be outdone by the navy, the air force upped the ante by contracting for a new swept-wing rocket plane of its own called the Bell X-2. With an improved rocket engine capable of propelling the X-2 to speeds of Mach 3, there were high hopes for this new program. Bell also increased the size of the original X-1 to allow its rocket engine to burn longer so that straight-wing design could potentially reach speeds as high as Mach 2.5. The problematic X-2 was delivered years late and crashed after only 13 powered flights over a 10-year period. Although it reached Mach 3.2 on its last mission, the airplane tumbled out of control after reaching that speed, killing its pilot, Capt. Milburn G. Apt.

The X-1A reached Mach 2.44, which was actually beyond its speed design envelope, and it, too, departed controlled flight in a series of wild gyrations that battered test pilot Chuck Yeager in the cockpit to the point where he said after the flight, "If I'd had an ejection seat. I would've used it." Several Bell X-1 aircraft and the first of two Bell X-2s were lost in mysterious explosions that turned out to be a reaction of Ulmer leather gaskets located within the fuel system to liquid oxygen. Despite those setbacks, these "first-generation" rocket ships certainly served their purpose in gathering valuable data about supersonic flight.

The competition for speed records between the army and the navy for the Pulitzer trophy that had ended in 1925 was now being reopened using experimental aircraft over the high desert of Southern California. These new speed marks were not officially recognized as official speed records by the FAI, however, but most of the public only read the headlines of what speeds were now being reached by the rocket ships, and the country at large was captivated by the news.

Stilettos, Bantams and Variable-Sweep Wings

Other X-planes were being built to investigate various types of wing shapes in the transonic and supersonic speed regime. While swept wings were optimum for the transonic speed range to Mach 1.2, higher supersonic speeds required a low-aspect wing with low

The most attractive airplanes of the early supersonic era: the Douglas Skyrocket and North American F-86 Sabre Jet. In this photo, the D-558-2 is now a glider having expended its rocket fuel. The Sabre chase plane is tucked-in tightly to provide visual assistance for the Skyrocket's pilot due to his limited field of view from the small cockpit windows. The Sabre pilot has deployed his speed brakes to stay in position with the gliding Skyrocket. (NASA via National Archives)

This talented trio is comprised of a civilian contractor test pilot, a U.S. Marine Corps combat veteran, and the most experienced NACA rocket pilot in history. From left, Douglas' celebrated test pilot Bill Bridgeman, USMC Col. Marion Carl, and NACA's Scott Crossfield, the first pilot to fly Mach 2. All three of these great airmen set their records in the Douglas-built research aircraft. (Wings & Airpower Historical Archive)

The last research aircraft to fly of the first-generation X-Planes was the trouble-plagued Bell X-2. Accomplishing its design objectives only in the last three months of a ten-year research program, the X-2 took high-speed and high-altitude flight to its absolute limits in the summer of 1956. Sadly, both of the X-2s built were lost in accidents after making 7 glide flights and 13 powered flights from 1952 to 1956. (Tony Landis Collection)

thickness-to-chord ratio as the optimum solution, and the wing sweep itself did not really matter. The super-sleek Douglas X-3 Stiletto was a project developed in cooperation between NACA and the air force to investigate supersonic use of higher-thrust turbojets, which hopefully could propel the X-3 to Mach 3 for a sufficient amount of time, allowing investigation of the "thermal barrier" where the skin temperature of an airplane could reach almost 600 degrees.

The Stiletto was built with a very long pointed fuselage to reduce profile drag and carry enough fuel for the jet to reach Mach 3. The test pilot had virtually no forward visibility and because of the shape of the fuselage had to eject downward rather than upward to escape the aircraft in an emergency. In fact, an elevator was necessary to winch-up the ejection seat, with the pilot in it, to get inside the airplane. The X-3 was the first aircraft equipped with an air conditioned cockpit, using a refrigeration unit to keep the pilot from being cooked at Mach 3.

Bell X-2 on lakebed after landing on its first glide flight on 27 June 1952. The Bell X-1 series had wheels because the original design was not originally envisioned for air launch. In contrast, the X-2 used a main landing skid with a conventional nosewheel, which collapsed during landings early in the program, causing the ground-looping aircraft to act like a high-speed plow. (USAF via Tony Landis)

Leaving a vapor trail from LOX propellant, the Bell X-2 is launched from its EB-50 mothership. The X-2 made only 13 powered flights but set new unofficial altitude and speed records of 126,200 feet and Mach-plus. On its final fatal flight, it reached Mach 3.2, but while attempting a high-Mach turn back to base at 70,000 feet, the aircraft departed controlled flight. Test pilot Mel Apt jettisoned the forward fuselage escape capsule from the aircraft but was unable to manually bail out from the falling capsule and open his parachute in time.

Bell X-1A (left) and X-1B were second-generation members of the Bell X-1 series. Jet fighter canopies replaced the fixed windshield of earlier X-1s, and the fuselage was extended for more fuel capacity to allow longer rocket burn times to exceed Mach 2. On 12 December 1953, Major Chuck Yeager flew the X-1A to Mach 2.44 (1,660 mph) and then safely landed the airplane after it spun out of control from inertia coupling, tumbling from a peak altitude of 75,000 feet. (USAF)

Unfortunately, the higher-thrust engines that were supposed to arrive never did, and the airplane was forced to use two smaller Westinghouse J34s that produced less than 5,000 pounds of thrust each in afterburner (versus 6,600 pounds each for the never-built J46). Because of this power deficit, the X-3 could not even achieve Mach 1 in level flight. It flew a total of 51 test flights, and as NACA test pilot Scott Crossfield reported, it became a Hangar Queen that was only trotted out covered by a tarpaulin to be shown to newspapermen as "a secret airplane that's the wave of the future."

The X-3 achieved one valuable aerodynamic discovery. During rapid rolls, the extremely long nose yawed violently. This is known as roll coupling divergence or the "dumbbell effect." The latter name was not meant to be derogatory: rather, the X-3's long, slim fuselage with heavy jet engines in the back, coupled with its short, stubby wings had an axis of rotation that rapidly deviated from the centerline of the fuselage when rolled rapidly. The rolling motion then coupled with a yawing motion and rapidly diverged. Another term for this phenomenon was "inertia coupling."

This problem surfaced on the air force's first production supersonic jet fighter, the North American F-100 Super Sabre, and led to several fatal crashes. Larger vertical tail surfaces were installed to dampen out the yaw divergence during rapid rolls. One further result of the X-3 experience was the small, low-aspect "double-diamond" trapezoidal wing, which was chosen by Lockheed Chief Engineer Kelly Johnson as the optimum shape for the wing of the Mach 2 F-104 Starfighter jet interceptor he was designing for the air force.

The Tailless Jet

Northrop's X-4 Bantam was another NACA/air force attempt to investigate the transonic speed regime using a tailless aircraft. Its configuration was similar to the Me-163 but used two turbojet engines rather than a rocket because it was only intended to investigate transonic speeds, and NACA wanted a flight duration of more than 3 or 4 minutes. The airplane was also quite small because its two turbojet engines produced only 1,600 pounds of thrust each, and to reach transonic speeds it had to be small and lightweight, unlike Northrop's other contemporary twin-jet aircraft, the heavyweight F-89 Scorpion interceptor.

The two X-4s built flew only a total of 92 flights over three years, but it confirmed that for jet aircraft approaching the sound barrier, the tailless configuration was not a good solution. The X-4's wing was swept back 40 degrees, but at Mach 0.88, the aircraft

Nicknamed "First of the Spacemen" after his record-breaking flight to 126,200 feet in the Bell X-2 on 7 September 1956, Capt. Iven C. Kincheloe is seen here posing with that aircraft for a PR photo at Edwards South Base. Note the blistered paint on the vertical stabilizer caused by excessive heating at speeds well above Mach 2. (Tony Landis Collection)

developed a severe pitch oscillation problem, which required some redesign to cure.

The pitch control surfaces for a tailless airplane are known as elevons and are located on the trailing edge of the wing. When a normal shock develops, the first control surfaces to be affected are the elevons and they lose some effectiveness at speeds less than supersonic. The tailless aircraft (also known as the flying wing) is very sensitive to center of gravity placement anyway, and the oscillating shockwave approaching Mach 1 has a dramatic effect on the controllability of the airplane.

One of the X-4's features was unique for the time, a fully irreversible hydraulic flight control system, which later became the

Rare photo of the X-3 taking off from Rogers Dry Lake at Edwards. Ground run was approximately 3 miles for the underpowered aircraft, and a new set of tires had to be used for every flight. The large grid on the side of the fuselage is for inflight photo calibration. (Tony Landis Collection)

The X-3 illustrating its trapezoidal wing intended for high Mach flight as well as the airplane's relatively small tail surfaces. The latter feature led to lateral directional instability called roll coupling when rolling at a high rate. NACA pilot Joe Walker flew the heavily instrumented X-3 to investigate this phenomenon after discovering it during a routine test flight, gathering much valuable data. (Douglas via National Archives)

norm for transonic and supersonic aircraft. The reason it had this advanced flight control system was that Northrop had developed it for its giant Flying Wing bombers, the XB-35 and YB-49, where they assumed the pilot would need physical assistance in moving the bomber's large, heavy elevons. In retrospect, it was good that the

The chief test pilot at the Air Force Flight Test Center in the mid-1950s was World War II and Korean War veteran Lt. Col. Frank K. "Pete" Everest, shown here in the cockpit of Bell X-2 Ship 2. Everest reached a record speed of Mach 2.87 (1,900 mph) on 23 July 1956.

Famed Douglas Aircraft Company test pilot Bill Bridgeman standing next to the X-3. He made 25 flights in that aircraft including its inaugural flight on 20 October 1952. After the successful contractor phase of its flight test program, the X-3 was turned over to the air force and NACA. (Douglas via the National Archives)

X-4 had this hydraulic elevon system, as un-boosted manual pitch controls in the transonic regime would have had a much worse effect on the flying qualities of the sensitive flying wing. Both X-4 airframes survived their flight test program and are on display in air force museums today, although most visitors have no idea why the X-4 was built.

Across the Atlantic in the immediate aftermath of World War II, an unofficial race developed between Britain and the United States to be the first to break the sound barrier, even if the aircraft had to be in a dive to reach that speed. The British developed a tailless aircraft called the de Havilland DH 108 Swallow, based on examination of the German Me-163. It used a 3,000-pound-thrust jet engine with more highly swept wings (43 degrees) to combat compressibility.

In September 1946, the British aviation world was shocked when test pilot Geoffrey de Havilland Jr., son of the legendary aircraft designer, died on a test flight while investigating the Swallow's high-speed handling characteristics. He entered a dive at Mach 0.91 at 10,000 feet and subsequently the wings failed catastrophically at higher speed. This may have been due to a normal shock forming on the trailing edge of the wing creating a "tuck under" steepening of the dive and subsequent overspeed.

A replacement aircraft was built with hydraulically boosted controls, and using this aircraft a new world speed record over a 100-km course of 605 mph was set in April 1948. That same year, John Derry, another British test pilot, inadvertently went supersonic (Mach 1.04) in a dive from 40,000 feet to 30,000 feet while he tried to control pitch oscillations. This aircraft was also later lost; in fact, three Swallows were built and all three crashed, killing their test pilots. It seemed that transonic airspeeds and tailless airplanes did not mix.

Swinging on a Spar

The Bell X-5 variable-geometry aircraft was another testbed to evaluate the effect of wing sweep on transonic flight performance.

Intended to achieve sustained high-Mach flight, the X-3 featured the first air-conditioned cockpit in an airplane. Dark tinted windows provided limited forward visibility, but this cockpit design protected the pilot with a heavy rigid structure rather than a clear Plexiglas canopy.

The Douglas X-3 Stiletto was long and pointed to minimize profile drag at high Mach, but was impractical as an operational jet aircraft due to limited visibility and marginal performance. Today, NASA is proposing a testbed low-sonic-boom demonstrator that actually has a long, pointed nose similar to the X-3 with the pilot relying on high-definition television displays in the cockpit for landing purposes. (Douglas via National Archives)

It was based on a German jet fighter, the Messerschmitt P 1101 captured at the end of the war, shortly before it would have first flown. Because Bell airplane's wing sweep could be changed in flight, the airplane's center of gravity was shifted, so it was necessary to devise a method for actually adjusting the X-5's center of gravity as well as the wing sweep angle. As a result, not only did the wing sweep change, but the entire wing assembly moved forward by means of an electric motor within the fuselage, sliding on a track as wing sweep increased or decreased.

Rather than modify the German prototype, Bell aircraft decided it would be simpler just to build their own airplanes based on the German design. Two X-5s were built and NACA conducted wind tunnel evaluations of the airplane at various sweep angles. It was noted in the spin tunnel that at 60-degree wing-sweep angle there might be problems recovering from a spin. Bell proposed adding a spin recovery parachute to the X-5, but both NACA and the air force decided it wouldn't be necessary.

The first aircraft flew in June 1951, and its variable-sweep wing was exercised successfully for the first time on the ninth flight, from 20 to 60 degrees and back. The X-5 was surprisingly successful, and from January 1952 to October 1955, the two airplanes logged 153 test flights. Unfortunately, the second X-5 was lost in a spin-recovery accident in October 1953, killing Air Force Test Pilot Capt. Ray Popson on his first flight in the airplane.

A major finding of the X-5 program was that rather than sliding the entire wing assembly forward on a track as wing sweep was increased, it was aerodynamically more efficient to move the wing pivot several feet outboard from the fuselage and merely transfer fuel to keep the center of gravity within limits as the wing was swept. This is basically the method that all variable-geometry aircraft use today, but the X-5 testbed pioneered the idea of sweeping wings in flight.

Combat aircraft in the 1960s and 1970s used variable-sweep wings, mainly to shorten takeoff and landing distances with minimal sweep and then transition to an intermediate sweep of 45 to 54 degrees for transonic flight. For supersonic cruise, the wings were fully swept back to more than 60 degrees. A major problem, however, was that by having the outboard wing pivots on both sides of the aircraft the wing-sweep mechanism's weight increased significantly, especially on larger aircraft. Boeing's winning entry for the U.S. supersonic transport (SST) was a variable-sweep goliath weighing more than 500,000 pounds, but once the actual design effort started, the wing-sweep mechanism's weight was so high the design was deemed impossible and the airplane was redesigned with a fixed wing. (Due to newfound environmental concerns and sheer cost, the SST was never built.)

One further flying testbed did not receive an X-plane designation, but proved to be a significant contributor to the shape of supersonic airplanes. Convair's XF-92 delta-wing jet became the proof-of-concept demonstrator for a high-speed jet interceptor. Originally planned to be powered by ramjets and rocket boosters and looking like something flown by Flash Gordon, the XF-92 was transformed into a more reasonable single-pilot jet fighter intended to explore the utility of a new type of wing, yet again based on German research work during World War II by aerodynamicist Dr. Alexander Lippisch (who subsequently came to the United States to work for Convair as a consultant). Lippisch had always been interested in a wing of triangular planform that looked like the Greek letter "Δ" (delta).

Such a wing would have a large area implying good low-speed performance, but in addition, the wing's aspect ratio was fairly low,

On the ramp at North American's plant in Inglewood, California, you can see the small tail on the early F-100As that had to be enlarged due to roll coupling. The Douglas X-3 discovered this phenomenon, but not in time for designers to realize that roll coupling could be a problem due to the swept wing and weight distribution of the modern jet fighter. (Wings & Airpower Historical Archive)

Increasing Jet Engine Thrust from Front and Back

For increased thrust for short periods of time, an afterburner (called reheat by the British) was developed to maximize the thrust coming out of the exhaust nozzle of the engine. Pouring raw fuel directly into the engine's exhaust stream further increased the velocity of the jet exhaust, but without increasing the temperature of the turbine. The afterburner definitely provided more engine thrust, but it was highly inefficient as the total engine fuel flow doubled, tripled, or even quadrupled to obtain only another 2,000 to 5,000 pounds of thrust. However, that figure was later improved with the advent of higher-thrust engines.

Original compressors in early jet engines were driven by a single turbine behind the combustion chamber, which was adequate for the early centrifugal-flow compressors with only one or two compressor stages. Development of the axial-flow engine used multiple compressor stages to increase the amount of airflow into the combustion chamber. This was not an optimal way to achieve the maximum compression possible. By using two separate turbines it was possible to have the initial low-pressure compressors moving at a different RPM than the high-pressure compressors, which were located just before entry into the combustion chambers. Thus, more air was packed into the combustion chambers, producing more thrust.

The two-spool engine, as it was called, could increase compression ratios from 12:1 to 25:1 on an engine roughly the same size as a single-spool engine, and the Pratt & Whitney J57 was one of the first two-spool engines. It could produce more than 10,000 pounds of thrust without using an afterburner and had much improved specific fuel consumption (MPG at cruise). More than 20,000 J57s were produced, used extensively in fighters, bombers, and even jet airliners, which generally did not go as fast as fighters, but could use the new speed and efficiency to fly longer ranges with the same amount of fuel.

By looking at the striking progression of absolute speed records from 1949 to 1955, it is obvious that afterburner-equipped aircraft set the pace for the increase of the speed record. Flying a 3-km course, the afterburner did not have to be engaged for very long, and the potent combination of swept wings and afterburner meant that low-altitude speed records rapidly approached the speed of sound. For fighter aircraft, afterburners seemed to be the ideal solution for aerial combat, which, although not lasting very long, allowed a distinctive advantage over a non-afterburner-equipped adversary.

Afterburners were also used to improve time-to-climb performance, as the atomic bomb became a major threat delivered by bombers, which generally flew at high altitude. So it was important that jet interceptors climbed to high altitude as quickly as possible. The navy's Douglas F4D-1 Skyray used its afterburner to set several time-to-climb records, such as 1½ minutes to 30,000 feet and 2½ minutes to 50,000 feet. Based on this stellar performance, the navy Skyray was drafted into the air force's air defense command to protect the continental United States against Soviet aerial attack.

Lockheed F-104A Starfighter with its General Electric J79 turbojet in full afterburner during an engine test run at night. The F-104 was capable of sustained flight at Mach 2, although it suffered from having limited range. The afterburner plume is especially impressive after dark; the dramatic increase in noise from any jet in afterburner is impressive at night or during the day. (USAF)

The predecessor to the F-104 in the air force interceptor role was Lockheed's F-94C Starfire, the ultimate evolution of the T-33 trainer, itself evolved from the original P-80 Shooting Star. The F-94C was the first air force aircraft equipped with an afterburner, which augmented the thrust of its Pratt & Whitney J48 turbojet by 2,500 pounds. However, with its straight wing, the F-94C was a subsonic aircraft at best. (Wings & Airpower Historical Archive)

The first Navy jet with an afterburner, and the first supersonic jet aircraft built by the Douglas Aircraft Company, the bat-wing XF4D-1 Skyray, was a variation on the theme of a pure delta-wing design and held numerous low-altitude speed and time-to-climb records. The production F4D-1 Skyray's Pratt & Whitney J57-P-8 turbojet produced 10,200 pounds of dry thrust and 16,000 pounds of thrust in full afterburner. (Wings & Airpower Historical Archive)

implying that it would have good high-speed performance. The long wing root would allow for stowage space for landing gear and fuel tanks adjacent to the fuselage, allowing greater structural strength of the wing. The delta-wing aircraft in its purest form would not have a horizontal tail but would instead use elevons on the trailing edge of the wing to control both roll and pitch. The removal of the horizontal tail would reduce drag and weight of the aircraft, making

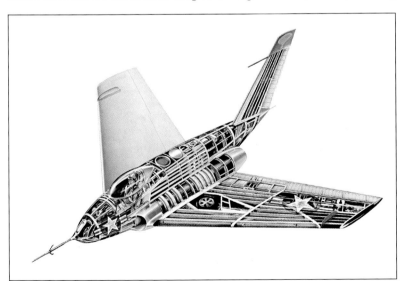

Cutaway illustration of the Northrop X-4 Bantam. The aircraft made its first flight at Edwards on 16 December 1948, and a total of two X-4s were built. Both aircraft survived their rigorous and probing flight test careers intact.

it possible for a delta-wing airplane to operate in the transonic realm with the limited-thrust turbojets of the postwar era.

The rather portly airplane was powered by a single J33 turbojet providing only 4,000 pounds of thrust. Test pilots all agreed it was underpowered for a fighter, but in reality, it was strictly a testbed for this new type of wing. Chuck Yeager was impressed when, on his second flight, he dove the airplane to Mach 1.05 without difficulty. When slowing to land he kept pulling the nose higher to slow down expecting to feel a buffet as he approached the stall, but he kept raising the nose higher in the air and the wing did not stall. Even more surprisingly, Yeager landed nearly 100 mph slower than Convair test pilots had recommended.

NACA's Scott Crossfield was the last pilot to fly the XF-92. He had never liked the flying characteristics of the airplane and on his final landing the nose gear collapsed and the airplane was officially retired. Nevertheless, although not officially an X-plane, as a testbed the XF-92 did an admirable job of exploring the characteristics of the delta wing. Both Convair and the entire aeronautical industry had taken note of the XF-92's success, and this new type of wing promised to be the key to future supersonic flight.

The World's First Jet Airliner Arrives

Meanwhile, across the Atlantic the British were also investigating high-speed flight. The Brabazon committee (headed by Lord Brabazon of Tara, who as British pilot number one famously flew "the first pig to fly" in 1909) was formed in 1943 to suggest what commercial aircraft should be developed to match the superior American

Northrop's X-4 was a U.S. Air Force/NACA attempt to investigate the transonic speed regime using a tailless aircraft. The airplane was quite small due to the fact that its two Westinghouse J30 turbojet engines produced only 1,600 pounds of thrust each. The X-4 had a 40-degree swept wing, but at Mach 0.88, the aircraft had an oscillation problem that resulted in a severe pitch-up and required some redesign in order to cure.

The Messerschmitt P 1011 had a variable-geometry wing to investigate the effect on critical Mach number. The war ended before it was completed but this prototype was used by Bell Aircraft Company to gather ideas for developing the Bell X-5 variable-sweep research aircraft, which resembles the P 1011. (USAF)

The Bell X-5 variable-sweep aircraft was based on the German jet prototype Messerschmitt P 1101, which had been captured at the end of the war, shortly before it would have first flown. Unlike the German design where the sweep could only be adjusted on the ground before flight, the sweep of the X-5's wings could be changed in flight. (NASA)

Convair XF-92 was a successful testbed for the concept of a delta-wing jet aircraft. Convair took note of its initial success, improved the breed, and built a series of legendary delta-wing fighters and bombers in the 1950s and 1960s that included the F-102 Delta Dagger and F-106 Delta Dart interceptors, and the four-engine Mach-2 B-58 Hustler. (USAF)

The X-5 investigated wing sweep, but also had to slide the wings forward and aft for controllability. Based upon this, the idea grew that for a variable sweep in-flight airplane, the wing pivot point should be placed outboard of the fuselage rather than in the center as on the X-5. This required two pivots to be installed rather than one, but it turned out to be a more efficient solution as illustrated in the diagram of the General Dynamics F-111 on page 157. (USAF)

Looking every bit the superlative test pilot he was, Chuck Yeager prepares to fly the experimental delta-wing Convair XF-92. Of his innate abilities as a test pilot, Yeager protégé, X-15 pilot, and Shuttle Astronaut Joe Engle once said, "Yeager doesn't just fly an airplane, he communicates with it." (Wings & Airpower Historical Archive)

Sporting a unique inversely tapered wing, Republic's novel XF-91 Thunderceptor was a good example of the innovative state of aircraft design in the post–World War II era, and became the first aircraft to achieve a speed of Mach 1 in level flight. Although never put into production, many of the XF-91's advanced features were incorporated into the F-105, Republic's member of the supersonic Century Series fighters. Note the four rocket tubes above and below the jet exhaust.

airliners from Douglas and Boeing. One of the commercial aircraft designs planned for the postwar era was to develop a jet airliner for use on routes covering the far-flung British Empire. The initial proposal was for an airplane carrying six passengers(!) and 2,000 pounds of mail with three jet engines.

This proposal was rejected, and instead a more traditional airliner with accommodations for 40 passengers powered by four Ghost 5,000-pound-thrust engines received the designation of de Havilland 106 Comet. A de Havilland Vampire jet fighter was modified with one of the Ghost engines to confirm operation at the high altitudes attainable with jet engines, and on 23 March 1948 de Havilland test pilot John Cunningham attained the first official jet absolute altitude record of 59,446 feet in the modified Vampire with a pressurized cabin.

The original Comet design was small and conservative with a wing sweep of 20 degrees, so flight testing proceeded well. The revolutionary airliner captivated the world, and eventually entered service in July 1952, well ahead of any other country. In its first year of service, nearly 28,000 passengers were carried. The Comet 1 flew at 40,000 feet carrying 36 passengers in comfort at speeds faster than a World War II Spitfire. Unfortunately, being first is not always best, as de Havilland soon found out.

The British aircraft industry suffered the penalty of being a pioneer in air travel as four Comet 1s were lost in a short period of time. Two of the accidents killed all the passengers and crew onboard after mysteriously plunging from the sky during climbs to high altitude.

The revolutionary de Havilland DH.106 Comet 1 jetliner entered passenger service in July 1952, well ahead of any other country, and carried nearly 28,000 passengers in its first year of operation. It flew 36 passengers at 40,000 feet in comfort, and at speeds faster than a World War II Spitfire. The jetliner had limited range, however, so many fuel stops were necessary to fly long distances, but high speed and the absence of vibration made the overall journey shorter and more pleasant. (Mike Machat)

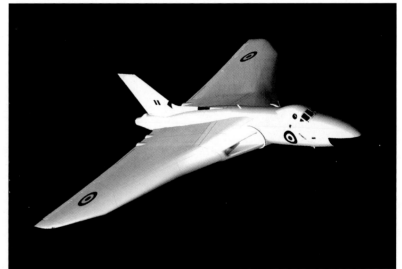

The first "B2" was actually this aircraft, one of three British "V-Bombers" developed to meet the Soviet Threat during the Cold War. The Avro Vulcan B.2 was a striking delta-wing design, and the largest delta-wing aircraft to fly until North American's XB-70. This 200,000-pound aircraft had a 111-foot wingspan and was powered by four 13,000-pound-thrust Olympus 104 turbojets giving the bomber a range in excess of 3,000 miles and a top speed of 630 mph.

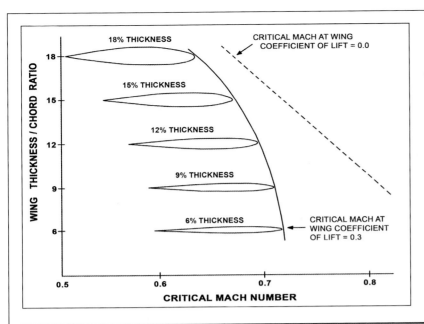

This wing thickness chart shows the interaction between the critical Mach number and the wing thickness ratio. Versions of the Bell X-1 rocket plane had different wing thickness ratios for test purposes. (Graphic digital enhancement by Craig Kaston)

Considered the ultimate jet fighter in 1948, a North American F-86A Sabre banks away from the camera plane showing its 35-degree wing sweep. First flown on 1 October 1947, the Sabre set the stage for development of the air-superiority fighter for decades. Powered by a 5,000-pound-thrust GE J47 turbojet, the F-86 could reach a top speed of 670 mph at sea level, and had an initial climb rate of 8,000 feet per minute. The Sabre achieved fame as a potent MiG killer during the Korean War, and established a world speed record in 1948. Test pilots Bob Hoover, Chuck Yeager, and "Pete" Everest each claimed the F-86 as their all-time favorite jet aircraft. (Wings & Airpower Historical Archive)

North American's F-86D Sabre interceptor. This was the first Sabre with an after-burner, which allowed significant increases in thrust for the high climb rates required to intercept incoming enemy aircraft. Speed records were set while approaching transonic speeds and it became obvious that to reduce shockwave drag it would be best to conduct record attempts on hot days when the speed of sound was high, to delay the onset of shockwaves. The optimum record attempt location would be a hot desert located at low altitude since jet engines produce the most thrust at low altitude due to higher air density. Hence, the record attempt shifted to the Salton Sea, which is near sea level versus Edwards, which is at 2,300-foot elevation. (USAF)

The absolute speed record exchanged hands five times in 1953; twice to the U.S. Air Force, twice to the British, and once to the U.S. Navy. The British Hawker Hunter, like the Sabre, had a swept wing and could reach supersonic speeds in a dive. Flown by legendary Squadron Leader Neville Duke, the Hunter set an absolute speed record of 741.5 mph on the 3-km speed course at Littlehampton, Sussex, England, on 31 August 1953. It also set a record of 709 mph over a 100-km closed-circuit course. (Wings & Airpower Historical Archive)

These accidents were due not to the jet engines, but to structural failure in the pressurized fuselage at 35,000 feet, leading to explosive decompression and subsequent destruction of the aircraft. The force of the explosion was so violent that despite the Comet's 500-mph speed, chips of paint from the upper fuselage were found imbedded in fragments of the outer wing.

All remaining Comet 1s were grounded, and the subsequent publicity and need to completely redesign the jetliner meant that the British lost their advantage in pioneering jet airliner service. But the Comet's brief two years of service showed that for commercial jets to be economical, aircraft had to be larger and have longer range to minimize refueling stops. These numerous flight pressure cycles could lead to metal fatigue and explosive decompression, as happened o the Comet 1. The Comet 1 had moderate wing sweep to be conservative, which limited its speed advantage over piston-powered airliners. Future jet airliners would have more wing sweep, more passenger capacity, and would carry more fuel as well.

The Avro Company, competing for a contract for both a jet airliner and jet bomber, came to the conclusion that a large delta-wing aircraft was the best solution for both designs. Avro got the contract to build a long-range jet bomber, but because no one had ever built a large delta aircraft, the company built four sub-scale fighter-size testbed aircraft known as the Avro 707 series. These aircraft were very successful in exploring both the low-speed and high-speed behavior of delta-wing aircraft.

Test results from the Avro 707 led to the success of the 111-foot wingspan Avro Vulcan jet bomber over the course of a 30-year operational career. The Vulcan's high subsonic cruise speed and large payload, combined with acceptable behavior in the low-speed range for landing, gave aviation designers a newfound confidence in using the delta wing on jet aircraft, even at supersonic speeds as subsequent British, French, Russian, and U.S. designs showed.

Official Swept-Wing Speed Records (1948–1956)

The Gloster Meteor and Lockheed P-80 set the first post–World War II jet speed records with their straight wings and confirmed the performance

Navy LCDR James B. Verdin and Douglas test pilot Bob Rahn exude the spirit of aviation in October 1953 at Edwards AFB. Rahn had just claimed the 100-km closed-course record of 728 mph in the Douglas XF4D-1 Skyray (background). Verdin had flown that aircraft to a low-level 3-km course record of 752.7 mph, making the Skyray the first carrier-based jet aircraft to hold the world's absolute speed record. (Wings & Airpower Historical Archive)

YF-100A setting the Speed Record for the last time on a required 100-meter altitude 3-km speed course. The dangers of flying a 3-km course in less than 10 seconds at an altitude of 330 feet above the ground at 750 mph for essentially a sporting event caused people to rethink future speed records. The rules were rewritten in 1954 and speed records could be set at any altitude using a 15- to 25-km course. The next speed record set by an F-100C flying at high-altitude at more than 800 mph was the world's first supersonic absolute speed record. All subsequent absolute speed records were supersonic at high-altitude. (USAF)

Blown tires and burning wheel brakes became more of a problem with the advent of the supersonic Century Series fighters and their faster landing speeds. The solution was to use a small-diameter ribboned parachute to slow the aircraft from its 150-mph touchdown speed to a more reasonable pace for wheelbrake use. Here, the production prototype North American YF-100A rolls out after landing and deploying its drag chute at Edwards. (The Museum of Flying)

Perhaps best known for his record-breaking flights in the rocket-powered Bell X-2, Lt. Col. Frank K. "Pete" Everest also set significant speed records in the North American YF-100A Super Sabre. Everest is seen here after his 754.9-mph dash at the low-level Salton Sea course on 29 October 1953. Two years later, AFFTC Commander Col. Horace Hanes upped the mark by flying a production F-100C to 822 mph, the world's first supersonic record flight. The F-100 had a 45-degree swept wing compared to the F-86's 35-degree wing sweep. (The Museum of Flying)

Preparing for high-G-load testing, Scott Crossfield is wearing the David Clark prototype full-pressure suit designed for the X-15, seated inside the U.S. Navy's Aero Medical Laboratory centrifuge in Johnsville, Pennsylvania, where a mockup X-15 cockpit was installed. Tests were to determine a pilot's ability to reach switches and controls while undergoing high-G forces. (Wings & Airpower Historical Archive)

An F-100 Super Sabre pilot prepares to board his jet wearing an anti-G suit, which was mandatory for withstanding the rigors of high-speed, high-performance flight in a supersonic aircraft. Pneumatically inflated under G-load, these "speed jeans" squeeze the pilot's legs and lower abdomen to keep his blood from pooling in his lower extremities, and away from his brain where lack of blood flow causes "tunnel vision," "gray out," and then "black out," or complete loss of consciousness, which can be potentially fatal especially at low altitudes. (The Museum of Flying)

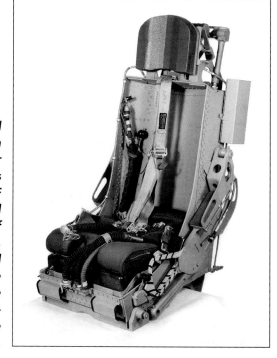

Close-up of an advanced version of the ejection seat. The small rectangular box at the upper right was a spring-loaded CHAFF dispenser that deployed hundreds of small strips of aluminum foil upon ejection, providing a briefly enhanced radar target showing the pilot's location for rescue purposes. The fiberglass-encased survival pack is visible just below the seat cushion.

This composite photo shows the evolution of a static ejection seat test, using a special ground-based test rig at Republic Aviation Corporation in Farmingdale, New York. The F-105 rocket seat with its test dummy pilot is shown ejecting from a pad, then activating the seat/man separator. The parachute deploys automatically and the "pilot" descends safely to the ground. Note that this is with zero altitude and zero forward speed, hence the term "zero-zero ejection seat." This entire sequence took less than 3 seconds when the seat was set in Mode 1 for takeoff and landing. (Republic Aviation Corporation)

superiority of the jet engine over a high-powered prop fighter. But now the swept wing made its debut. The air force attempted to set a new world speed record at the Cleveland National Air Races in 1947 using the new North American F-86A Sabre with a 5,200-pound-thrust General Electric J47 jet engine flown by Maj. Robert Johnson in front of 80,000 spectators.

Unfortunately, timing camera breakdowns prevented the record from being officially certified, although the average speed in the timed runs was 669 mph. Another record attempt was made at Muroc in front of a considerably smaller crowd and on 15 September 1948, Johnson set a new world record of 671 mph. One thing notable was that the aircraft was a standard production Sabre with full fuel and ammunition rather than a specially modified speed racer like the P-80R.

The Sabre's record didn't stand for very long, though, as the new F-86D interceptor model of the Sabre jet was being tested at Muroc.

Control and Stability of Fixed-Wing Airplanes

One of the greatest contributions of the early aviation pioneers was the development of an approach to the control and stability of an airplane. To remain level required a balancing act between the lift force and the weight force. To control the attitude of the airplane it was necessary to have a means of generating a lift to counteract the pitching motion. This was usually in the form of a horizontal tail surface with a hinged portion (elevator), which was itself also a wing and which was controlled by the pilot to command the pitch attitude of the airplane.

The Wright brothers also discovered that the way to turn an airplane was to bank the wing and allow the large lift vector tilt to cause the airplane to turn. They originally used differential wing warping to cause this tilt, although later control surfaces somewhat similar to the elevator were mounted on the trailing edge of the wing and they were used to tilt (or bank) the airplane. A rudder was mounted behind the wing to point the nose of the airplane into the relative wind in the direction in which the vehicle was turning.

The rudder itself acted similarly to the feathers on an arrow to control the yaw (left/right direction), and the Wright brothers actually interconnected the rudder to the wing warping to ensure a coordinated turn with no sideslip (angle between the x-axis of the airplane and the relative wind of its motion through the air). Sideslip greatly increases the drag on an airplane and with the limited power of the early engines, it could cause the airplane to slow down to the point where it was uncontrollable and crashed.

The stability of an airplane refers to its tendency when disturbed to return to its original position. *Static stability* refers to the tendency to return to the original position, whereas *dynamic stability* refers to the motion the airplane performs while it is returning to the original position. *Positive stability* means the system always returns to the original position. *Negative stability* means the system does not return to the original position by itself and some active controls are necessary to drive it to the original position. *Neutral stability* means when moved to a new position the system remains in that position and does not return to original condition.

Dynamic stability, the motion the airplane exhibits while returning to the original condition, is usually an oscillatory motion that often overshoots the original position and then has to reverse the velocity to go back to the original condition with several overshoots being common. *Positive dynamic stability* indicates that only a few overshoots occur while *negative dynamic stability* means that the overshoots continue to the point that the oscillation never damps out and may actually diverge (grow larger with time).

Stability in the longitudinal (pitch) axis is represented by phugoid and short-period motion. The latter is similar to driving a pickup truck at high-speed over a washboard dirt road while the former is closer to a rollercoaster ride. Directional (yaw) and lateral (roll attitude) motion are coupled in an airplane and stability is measured by disturbing the airplane from a stable condition and noting the subsequent yaw and roll behavior. Swept-wing airplanes, especially, exhibit a so-called Dutch roll combination (the name comes apparently from standing behind a speed skater from Holland and observing the way arms and rear end move in opposing harmony). So-called *spiral stability* also measures how an airplane in a bank will roll itself out of the bank (hopefully) if left to its own devices.

In the 1920s, inventor and pilot Lawrence Sperry developed a rudimentary automatic pilot using gyroscopes that allowed the pilot to take his hands off the controls for extended periods of time because the gyros automatically sensed any deviation from the original attitude and automatically adjusted the control or trim surfaces to return to the original attitude. This is the basis of more modern stability augmentation systems (SAS), which can be used to augment stability of designs with inherently poor hands-off stability.

During World War II bombing missions the bombsight was used to send correction signals to the autopilot so that the bombardier was actually steering the airplane by sighting on the target. Stability augmentation systems and this more complex use of the autopilot were forerunners to the sophisticated flight control system of today with digital computers that are used to control the flight path of the airplane and allow designs to offer "carefree handling qualities" to the pilot.

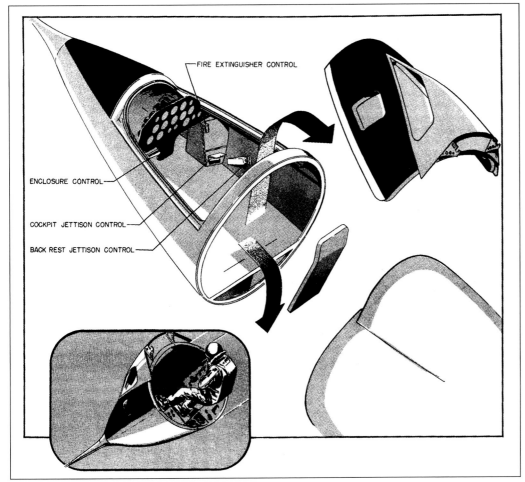

FIRE EXTINGUISHER CONTROL

ENCLOSURE CONTROL

COCKPIT JETTISON CONTROL

BACK REST JETTISON CONTROL

Douglas tech art showing how pilot could leave the Skyrocket's forward fuselage after jettisoning it from the airplane to allow bail out. In actuality, the pilot would find it much more difficult than shown here, as the unstable forward fuselage would more than likely be tumbling. A similar escape method was used for the Bell X-2 on its last flight, but the pilot was unable to egress the capsule in time. (Douglas Aircraft Company)

hot environment to reduce Mach number and avoid transonic shockwave drag. The Libyan Desert was even hotter than the Salton Sea, and the pilot and some of the aircraft systems had to be packed in ice prior to engine start because the aircraft cockpit cooling system was woefully inadequate to cope with the oppressive temperature on the ground.

Not to be outdone, the U.S. Navy entered the fray with its new delta-wing Douglas XF4D-1 Skyray interceptor, also equipped with an afterburner to allow fast climbs after being launched from an aircraft carrier. Up until this time, navy jet fighters had been equipped with straight wings to allow them to safely land on carriers at slower speeds. The Skyray, with its revolutionary delta wing, promised high speeds as well as low speeds at high angles of attack on carrier approaches, and the aircraft was being tested at Edwards Air Force Base in 1953. The navy task force proceeded to nearby Salton Sea, as did the F-86D team, and raised the 3-km speed record to 753 mph on 3 October 1953.

Whereas the first swept-wing speed records were set by the Sabre, the larger and more powerful F-100 Super Sabre was now in flight test using essentially the same afterburning J57 as the navy F4D. The prototype YF-100A exceeded the speed of sound in level flight on its first two test flights and set a new absolute speed record over the Salton Sea of 755 mph on 29 October 1953.

This record was significant because it was the last time the absolute speed record was set using a 3-km course at 100-m altitude. The absolute world record was now approaching supersonic speed where there were known problems with controllability and possible pitch-up, and FAI rules were revised in 1954 so that a speed record could be set by making two runs in opposite directions at any altitude over a 15- to 25-km course. These new rules also stipulated that the altitude could not deviate more than 50 m during the measured portion of the record run to avoid using a dive to increase speed.

New photo theodolite cameras were used to photograph and track the airplane on the speed course through the gates. On 20 August

Flying over a 3-km course at the Salton Sea near San Diego, California, Capt. Jason Wade Nash of Edwards set a new mark of 699 mph. This record did not last very long either, as Lt. Col. William Barnes set another new record of 760 mph on the same Salton Sea course. The difference was not only in having an improved afterburner, but also the fact that the new record was set at a temperature of 105 degrees F versus 76 degrees F for the earlier record. At 105 degrees F, the speed of sound was 797 mph, so although the airplane was traveling faster, its Mach number was lower than the earlier record.

The F-86D's record was quickly broken by a British Hawker Hunter. Like the Sabre, it was a swept-wing airplane that could achieve supersonic speed in a dive. Using this afterburner-equipped aircraft, which was painted a bright red and had a more pointed nose than the standard Hunter, squadron leader Neville Duke upped the speed record to 728 mph on the 3-km speed course at Littlehampton, England. He also set a record of 709 mph over a 100-km closed circuit on 7 September 1953.

The Hunter's record fell after only 18 days when a swept-wing Supermarine Swift fighter set a new record of 735 mph in the Libyan desert using the same tactic as the F-86D, low altitude in a very

The entire nose section of the Bell X-2 served as an escape capsule for emergency bailout. Sadly, the unsuccessful use of this device in 1956 became the compelling factor that led engineers to the realization that a simple, strong, and high-powered ejection seat was the best option for escaping a high-performance aircraft in an emergency. This was borne out when seats were fitted in the hypersonic X-15, Mach 3.2 Lockheed SR-71, and even the first space shuttle for use below 100,000 feet. (Tony Landis Collection)

Showing how well the nose escape capsule integrated with the overall airframe, Bell X-2 Ship 2 poses on the cold winter ramp at Bell's Niagara Falls facility in 1952. (Wings & Airpower Historical Archive)

Douglas X-3 illustrating the downward-ejection seat necessary because of the fixed triangular cross-section cockpit. Early Lockheed F-104s (top photo) also had downward-ejection seats, which were responsible for the deaths of several test pilots when the General Electric J79 jet engine quit immediately after takeoff. The downward-ejection seat velocity added to the already low altitude and was a death sentence for pilots. (Douglas via National Archives)

1955, the first production F-100C was flown by Col. Horace Haynes from Palmdale, California, south toward Los Angeles at high altitude to set the new absolute speed record of 822 mph, thus becoming the first supersonic speed record. People's reaction to the sonic booms over Pasadena were not recorded, but the era of supersonic speed records had firmly arrived. This potent combination of more powerful jet engines with afterburners and swept wings meant the sound barrier was no longer a speed limit. Operational jet fighters could now exceed speeds that a few years earlier had been deemed impossible.

Escape Systems for High-Speed Aircraft

As aircraft speeds increased, it became nearly impossible for a pilot to leave the airplane in the hurricane-force airspeeds of more than 200 mph. Once free of the cockpit, it also became nearly impossible to avoid hitting the airplane's tail. Ernst Heinkel installed primitive ejection seats in his early jet designs and even some prop airplanes. The Bell X-1 rocket plane had no ejection seat, but had a side cockpit door that was not really a satisfactory means of escape. Designers of both the Douglas D-558-1 Skystreak and D-558-2 Skyrocket envisioned releasing the entire forward fuselage from the rest of the aircraft with the pilot still in it. Once the forward fuselage was clear, the pilot then manually bailed-out of the free-falling cockpit, which hopefully wasn't tumbling out of control.

As ejection seats improved and survivability increased, this fuselage separation technique fell out of favor, although the swept-wing Bell X-2 rocket plane still employed that concept for high-speed escape. The use of the X-2's escape capsule unfortunately led to the death of air force test pilot Capt. Milburn G. "Mel" Apt when he was unable to manually open his parachute after separating the X-2's forward fuselage from his out-of-control airplane after reaching Mach 3 at 70,000 feet. The German approach of ejecting a pilot's seat out of the crippled airplane as rapidly as possible ultimately proved to be the best possible solution, and has stood the test of time as seats became far more advanced over the years.

Human body limitations under vertical acceleration in a seated position for short periods of time (less than a second) are approximately 20 Gs without restraint and pre-positioning, and 25 Gs if securely strapped to the seat, with the human spinal column being the weak link. Several jet aircraft used downward-firing ejection seats, notably the Lockheed F-104 Starfighter, which had a T-tail, and Douglas' X-3 Stiletto, which had an oddly shaped enclosed cockpit that made upward ejection impossible.

The downward-firing ejection seat was considered a death trap for low-altitude bailouts and was quickly rejected by the pilot community. Famed Edwards' test pilot Capt. Iven Kincheloe, known as "First of the Spacemen" for his Bell X-2 altitude record of 126,200 feet in September 1956, was killed in a non-survivable bailout from an F-104A after its J79 engine flamed-out shortly after takeoff. After rolling the jet inverted and ejecting upward, Kincheloe landed in the fireball of the crashing Starfighter. Today, only one aircraft still has downward ejection seats: the venerable 60-year-old B-52 Stratofortress, with the hapless seat occupants being navigators who sit in the "basement" of the two-story flight deck rather than upstairs with the pilots, who eject upward.

As aircraft speeds increased, it became necessary to develop safe and reliable escape systems for use at very low altitudes, even immediately after takeoff. The development of rocket motors provided more tai-lored thrust than an explosive cannon shell, and the use of a small explosive charge to reduce time for parachute deployment allowed successful escape at zero altitude and zero airspeed while remaining within human physiological limits. Once free of the aircraft a small drogue chute deployed from the seat to stabilize and slow it.

This deceleration was necessary to keep nylon parachute canopies from ripping when opened at high speeds, and to reduce the extreme physical distress to crewmembers caused by the violent opening shock. With the seat stabilized, the seat-man separator system automatically unlocked all belts and harnesses, and in effect, threw the pilot out of the seat. At high altitudes, the pilot free-fell down to breathable air where his parachute opened automatically, governed by a barometric sensor. At low altitudes, the chute opened less than two seconds after seat/man separation.

Not all aircraft ejections occur at high speeds, however, and for today's jet pilots, the low-speed corner of the flight envelope can be just as dangerous. High-performance "zero-zero" ejection seats such as the U.S.-designed ACES II or advanced British Martin-Baker seats can eject from an inverted aircraft as low as 500 feet and still save the pilot. Martin Baker seats have also been used successfully since the early 1950s on carrier-based U.S. Navy aircraft, once even ejecting a pilot from his disabled jet after landing on a carrier, and depositing him safely right on the flight deck.

Although downward-ejection seats were death traps at extremely low altitudes, the bombardier/navigator in the Boeing B-47 really had no other option. Seen here in tests conducted at Eglin AFB, Florida, a live test subject is ejected out of a B-47 to determine the seat's survivability. Other aircraft fitted with downward-firing ejection seats were the Douglas X-3 and early Lockheed F-104 Starfighters. (Wings & Airpower Historical Archive)

This frame from an often-seen, but highly graphic, photo sequence shows the pilot of a Vought F8U-1 Crusader ejecting from his stricken craft as it plunges off the end of the angle deck on the USS Franklin D. Roosevelt (CVB-42). This aircraft from VF-11 Red Rippers had sheared off a main landing gear strut, but the pilot survived to fly another day. (U.S. Navy)

Demonstrating how aerospace structural design can ensure pilot survivability, this Republic F-84F flown by the Belgian Air Force survived a mid-air collision and was able to land despite having its entire nose section sheared off. Many aerospace structural safety design features were incorporated into automotive design beginning in the 1970s, which noticeably reduced auto fatality rates.

Chapter 5

JET AIRLINERS AND MACH 2 FIGHTERS (1954–1962)

The world's largest and most powerful single-engine, single-seat aircraft in 1957 was Republic's Mach 2 F-105 Thunderchief. Developed to deliver a single nuclear store to an adversary during the height of the Cold War, this fighter-bomber delivered 75 percent of all the ordnance dropped over North Vietnam. The F-105 went supersonic on its maiden flight, and held the 100-km closed-course speed record of 1,216.5 mph set in December 1959. (Mike Machat Collection)

New jet aircraft developed in the early 1950s were a far cry from those of the previous subsonic "blow-torch era," and advancements in both aeronautical design and aircraft performance were made by leaps and bounds. In 1956 alone, more new aircraft made their first flights than at any time since World War II, with many military types going supersonic on their maiden flights. In a bizarre twist, although this would never be allowed today, aircraft manufacturers were actually paid large bonuses for their new jets exceeding Mach 1 on a first flight.

The Jet Absolute Speed Record Exceeds 1,000 mph

The British continued their interest in the delta wing from an experimental standpoint, using the experimental Avro 707 as a subscale flying prototype of the Vulcan jet bomber. The first testbed was the Fairey Delta FD 1, which first flew in 1952. It did not fly particularly well, however, and was actually grounded for two years, having been deemed a "dangerous aircraft." It was destroyed in an emergency landing in 1956 after its two-year hiatus on the ground and was not particularly successful except to show what not to do on the Fairey Delta FD 2.

Whereas the FD 1 almost looked like a rotund caricature of a delta-wing airplane, the Fairey Delta FD 2 looked like a pure thoroughbred. Powered by a slim Avon axial-flow jet engine producing 10,500 pounds thrust "dry" and 14,500 pounds with afterburner (or reheat), the sleek airplane looked like something out of a Buck Rogers comic strip. It had Britain's first hydraulically powered flight control system and could easily go supersonic at altitude with no buffeting or vibration at all. On 10 March 1956, legendary British test pilot Peter Twiss set the world's first absolute speed record over 1,000 mph flying 1,132 mph at 38,000 feet. This translated to a speed of Mach 1.72. The aircraft was also the first to have a movable "droop nose," which gave pilots a better view of the runway while flying an approach at the delta's characteristic high angle of incidence on landing. This aircraft is

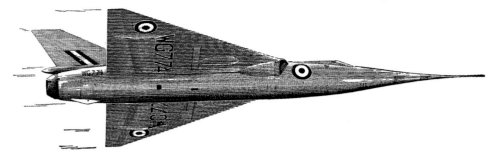

Only seven months after the F-100 set speed records over the low-level Salton Sea course, Great Britain once again entered the fray with the stunning new delta-wing Fairey Delta 2. Making two passes over a 9-mile course at 38,000 feet, famed Fairey test pilot Peter Twiss achieved a combined speed of 1,131.76 mph, the world's first record over 1,000 mph. This ad proudly touting that achievement appeared in Time magazine in April 1956. (Mike Machat Collection)

Boosted Flight Control Systems

In the early days of aviation, the pilot's control stick was connected via wires or rods to control surfaces such as the elevator, ailerons, and rudder. This was known as a reversible flight control system because the control surface "pushed back" at the pilot, giving him feedback on controlling the airplane. As aircraft speeds increased it became harder for the pilot to physically move the control surfaces due to increased relative wind velocity. Various tricks were used to give the pilot mechanical advantage in moving the controls, including having the control stick actually connected to small "tab" surfaces that caused the control surfaces to move.

Nevertheless, as aircraft became larger and the jet engine arrived, aircraft could move through the air at speeds of more than 400 mph. It became obvious something would have to be done to assist the pilot in moving the control surfaces. The answer was hydraulics, which were used to raise and lower the landing gear, meaning hydraulic lines were readily available.

Initially, elevators and ailerons were hydraulically boosted similar to power steering on an automobile; if hydraulics failed, the pilot could still fly the airplane but with heavier control forces and within a limited speed envelope. Later the pilot's control stick was connected to hydraulic motors, which in turn connected to control surfaces to move them. These were known as irreversible flight control systems, because the pilot did not have any direct feedback, and could not "feel" the control forces as he flew the airplane at different speeds. Various methods were devised to provide artificial feel to the pilot with systems involving springs and bob weights connected to the control stick to make control forces heavier at higher speeds and higher G levels during tight turns in aerial combat.

The Northrop N9M flying wing testbed in 1944 had an early irreversible flight control system because it was the testbed for the new 172-foot-wingspan XB-35 Flying Wing bomber. The XB-35 was so large it needed hydraulic assist for the flight crew to move the control surfaces. The XB-35 never entered production, but the N9M is still flying today on the air show circuit, and the pilot who flies it comments that although the 65-foot airplane flies very nicely without hydraulic boost, he always uses it for landing because the airplane suffers from control reversal if the landing speed is too slow and the angle of attack is too high.

This was another reason for having hydraulic flight controls on swept-wing jet aircraft; they could suffer control reversal and even pitch-up problems, especially in the landing pattern. Today, virtually all high-speed jet aircraft have irreversible flight control systems, although 21st-Century aircraft are beginning to use electric motors rather than hydraulic motors to move their flight control surfaces.

If there were ever an airplane simply too far ahead of its time, it would have to be the futuristic Northrop XB-35 flying wing (shown) and its jet-powered YB-49 brethren. This 172-foot-wingspan goliath was designed, built, and flown using mid- to late 1940s technology, and ultimately suffered from a lack of such advanced technologies as a Stability Augmentation System (SAS) and Digital Flight Control System (DFC) both of which were employed successfully on its digital-age successor, the Northrop B-2 Spirit Stealth Bomber. (Wings & Airpower Historical Archive)

McDonnell's brutish F-101 Voodoo was originally envisioned as a long-range penetration fighter for escorting B-52s. It subsequently changed roles becoming a nuclear strike aircraft, an interceptor, and a reconnaissance jet (shown). Although the airplane was large and heavy, being equipped with two afterburning Pratt & Whitney J57 turbojets allowed it to re-capture the speed record from Britain with a run of 1,208 mph (Mach 1.82). For the remainder of the Cold War, absolute speed records continually changed hands between the USAF, U.S. Navy, and the Soviet Air Force. (Mike Machat)

proudly displayed today in a British museum as a milestone of flight and also the last British jet aircraft to hold the absolute speed record.

Not to be outdone, the U.S. Air Force set out to reclaim the absolute speed record and succeeded on 12 December 1957, when a McDonnell F-101A Voodoo powered by two Pratt & Whitney J57 turbojets of 14,500 pounds static thrust each with afterburner set a new record of 1,208 mph, which translated to Mach 1.82. The Voodoo was originally intended to be a long-range bomber escort for the B-52, but because it had insufficient range for that role, it switched to a fighter bomber and interceptor, the latter mission requiring two powerful afterburner-equipped engines to climb to altitude and intercept high-flying Soviet bombers.

The Jet Bomber Arrives

After World War II, turbojet engines propelled fighters to speeds approaching Mach 1, or the speed of sound. Military aviators wanted to take advantage of the jet engine for bombers as well but the relative inefficiency and fuel consumption of early jet power plants presented problems for long-range bombers. To achieve long range, they would have to carry so much fuel that they'd be unable to carry heavy nuclear weapons. Medium-size bombers developed during World War II generally had two engines with less range than heavy bombers, and were relatively smaller. Initial jet designs tended to be medium bombers with twin nacelles housing four to six relatively low-thrust engines. The North American B-45 Tornado was America's first operational jet bomber, and although not very successful with its straight wings and limited range, it gave air force bomber crews initial experience with turbojets.

North American's subsonic B-45 Tornado was America's first operational jet-powered bomber. Although not very successful as a high-speed strategic aircraft with its straight wings and limited range, it nevertheless gave USAF bomber forces their initial experience with jet engines. (USAF)

Boeing's revolutionary swept-wing B-47 Stratojet first flew in 1947, and set the precedent for both military bombers and commercial jet airliners for the next several decades. Capable of speeds in excess of 600 mph, the six-engine Stratojet was built to carry atomic bombs to distant targets, but not too distant as its range with a 10,000-pound payload was 4,000 miles. The problem was how to extend the B-47's range; in-flight refueling became the solution.

The airplane that wouldn't die, Boeing's irreplaceable B-52 Stratofortress. Shown here taking off from Boeing Field in Seattle armed with two nuclear-armed AGM-28 Hound Dog missiles, the B-52 was capable of long-range, long-endurance missions anywhere around the world using aerial refueling. Powered by eight Pratt & Whitney J57s, "tall-tail" B-52s flew around-the-clock alert missions ready to launch a retaliatory strike on any potential foe. The copious amounts of black smoke was from power-boosting water injection used on takeoff. (USAF)

Several other bomber concepts were built, all suffering from the limitations of straight wings as well as poor fuel efficiency. In 1947, a radical new jet aircraft from the Boeing Company impacted not only future bomber designs but even airliners to come. It was the B-47 Stratojet with very thin 37-degree swept wings, powered by six General Electric J47 turbojets with 5,000 pounds of thrust each (7,000 pounds with water injection for takeoff). These engines were housed in pods suspended under the wings. Originally done for fire safety reasons, the weight of the pods actually kept the thin wings from flexing too much at the high speeds the B-47 flew, helping solve the high-speed flutter problem associated with thin wings. The B-47 was designed to carry nuclear weapons, which were quite large and heavy (7,600 pounds for the Mark 6 bomb), but with its single

On New Year's Day 1929, a U.S. Army Fokker C-2 transport named The Question Mark took off from Los Angeles airport and stayed airborne for 151 hours, receiving food and fuel from two Douglas C-1 tanker planes using a fuel hose and cable via a hatch in the top of the airplane. Thus, military aerial refueling began in America. (Douglas Aircraft Company)

Boeing KC-97 tankers used the flying-boom refueling system and could transfer fuel at the rate of 5,000 to 6,000 pounds-per-minute. They were modified versions of the Boeing 377 Stratocruiser developed from the B-50, but because they were slow piston-powered aircraft, the new swept-wing air force jet fighters and bombers had to slow down to uncomfortably low speeds while they took on fuel. To compensate, the tankers were "tobogganing" downward as fast as they could possibly fly, and at full power as well. (Mike Machat)

bomb bay, could carry a 20,000-pound payload for a combat radius of 2,000 miles at 40,000 feet, and at an impressive cruise speed of 550 mph.

The aircraft had a three-man crew and more than 2,000 of them were built despite the fact that they did not have the range to fly combat missions from the United States to Moscow. As revolutionary as it was, the B-47 was actually a stop-gap measure until the mammoth eight-engine B-52 intercontinental-range bomber could be developed. In the interim, the air force realized the problems of limited range for the early jets and explored the possibility of refueling in midair to extend their range.

Aerial Refueling Becomes Routine for Military Aircraft

Transferring fuel inflight was not exactly a new concept in the early years of aviation. On New Year's Day 1929, a U.S. Army Fokker C-2 transport named *The Question Mark* took off from Los Angeles airport and stayed airborne for 151 hours, receiving food and fuel from two Douglas C-1 tanker biplanes using a fuel hose and cable via

View from the KC-97 boom operator's window refueling a B-47. The flying-boom method required that a receptacle be installed on the top of air force receiver airplanes and that a skilled boom operator looking through a window be in the belly of the tanker to insert the boom into the receptacle via controls at his station. (USAF)

The jet-powered Boeing KC-135A Stratotanker first flew at Seattle in August 1956, and represented a quantum leap in aerial refueling technology. The large air force order for the KC-135 gave the airlines confidence that the Boeing 707 would be a success and led eventually to a virtual explosion in jet airliner traffic. Meanwhile, the B-52 bomber was given nearly unlimited range by using jet tankers. (USAF)

Boeing's perfect match: a KC-135A powered by J57 turbojets refuels a fellow KC-135B fanjet-powered Stratotanker with Pratt & Whitney TF-33s, and both airplanes traveling at normal jet cruise speeds. Max gross takeoff weight of the KC-135A was 297,000 pounds and ferry range on its 30,000-gallon internal fuel load was approximately 5,000 miles. Today, KC-135s have been re-engined with more efficient 22,000-pound-thrust SNECMA CFM-56 high-bypass ratio turbofans and have upgraded digital cockpits as well. (The Museum of Flying)

a hatch in the top of the airplane. *The Question Mark* flew along the California coast between L.A. and San Diego, and its crew included two future chiefs of staff of the army air forces. All crewmembers received Distinguished Flying Crosses while the pilots of the tanker airplanes who had taken off and landed day and night 43 times, often in terrible weather, received only letters of commendation. This started a long tradition in the air force of never fully appreciating tanker crews until you realize you're about to run out of gas.

The British had been exploring aerial refueling since 1930 to be able to fly passenger planes to far destinations such as India, South Africa, and across the Atlantic to Canada and the United States. In 1932, Sir Alan Cobham began investigating the problem using large twin-engine Handley Page aircraft converted as tankers. These tankers trailed a hose behind them in flight and the receiver aircraft then approached and inserted a pipe mounted on the receiver aircraft into a funnel at the end of the tanker hose, maintaining formation while fuel was pumped. This method, later known as the "probe and drogue" method, is still in use today. Its one disadvantage was that the flexible, long hose imposed limitations on the fuel flow rate, making it possible to transfer only approximately 2,000 pounds of fuel per minute.

Receiver aircraft, a two-seat Lockheed F-16D, pulls into position behind the boom of the Boeing KC-135B Stratotanker before making contact and taking on fuel. Note the green, yellow, and red signal stripes painted on the boom's extendable sections as a visual guideline for the refueling aircraft. Today, it is estimated that somewhere around the world, an aircraft is being refueled inflight every 52 seconds. (Mike Machat)

As a result, the U.S. Air Force elected to develop a "flying boom" that had a rigid metal boom with small stabilizing fins installed below the tail of the tanker with a short hose that telescoped in and out of the boom. The boom operator, while looking out a window in the underside of the tanker, manipulated the fins on the boom to insert the end of the hose into a receptacle atop the receiver aircraft. He monitored the receiver aircraft's position to ensure safe transfer of the fuel, disconnecting the hose if the receiver moved too far out of position. Because the hose was short, it was possible to transfer fuel at an initial rate of 6,000 to 8,000 pounds per minute using more powerful pumps on the tanker aircraft. Boeing KC-97 tankers

(modified versions of the Boeing Stratocruiser developed from the B-50) used this method, but because they were slow, piston-powered airplanes, the faster swept-wing air force jet bombers and fighters had to slow down to uncomfortably low speeds while they took on fuel. Jet engines were installed in pods under the wings of the tankers to boost their speed, but it was still not an optimum solution.

This composite photo shows a map of the United States super-imposed on a California Air National Guard F-86A at Van Nuys Airport north of Los Angeles. Aptly named California Boomerang, this Sabre was flown from Van Nuys to McGuire AFB, New Jersey, and back again, all during daylight hours! This flight, made in 1955, gave U.S. airlines a very compelling preview of what jet speeds could do for commercial airliners.

If you recognize the face of this young airman sitting in the cockpit of the California Boomerang F-86A, it's because he is Jack Conroy, father of the "Super Guppy" family of converted Boeing Strato-cruisers. These giant transports were the world's largest-capacity cargo aircraft at the time, and were built by Aero Spacelines, which was founded by Conroy.

Five years before rocketing into space to become the first American astronaut to orbit the Earth in his Mercury capsule, Marine Col. John Glenn rocketed into the record books flying this Vought F8U-1P photo-recon Crusader coast to coast in only 3 hours and 23 minutes. The transcontinental speed dash, called Project Bullet, was flown from NAS Los Alamitos, California, to Floyd Bennett Field in Brooklyn on 16 July 1957. (Wings & Airpower Historical Archive)

A fleet of USAF McDonnell RF-101 Voodoos prepares for Operation Sun Run in the first use of the air force's new Boeing KC-135 jet tankers for a transcontinental speed dash. The Voodoos flew from California to New York and then back to California in just under seven hours, although their one-way trip time was not hugely different than that clocked by John Glenn's Vought F8U-1P Crusader, which had to slow down to refuel from piston-powered tankers. (USAF)

The U.S. Navy adopted the probe and drogue system for their jet fighters in 1954, but the air force chose the flying boom method due to the necessity of refueling large bomber aircraft taking on more than 100,000 pounds of fuel. Air force aircraft, including fighters, thus had a refueling receptacle installed on the top of the fuselage while navy aircraft used a refueling probe mounted on the forward fuselage. In 1956 the first KC-135 Stratotanker, similar to the Boeing 707 jetliner, was introduced using the flying boom method and replaced the KC-97s. Using a fleet of KC-135 tankers on 16–17 January 1957, three B-52s circled the earth nonstop in 45 hours and 19 minutes with the bombers refueling only three times during the flight. On 10–11 January 1962, a B-52H (powered by eight Pratt & Whitney TF33 Turbofan engines of 17,000 pounds of thrust each) reclaimed the absolute record for distance in a straight line set by the U.S. Navy propeller-driven P2V Neptune Truculent Turtle in 1946 by flying unrefueled from Okinawa to Madrid, a staggering distance of 11,333 miles.

Aerial refueling thus extended the range of military jet aircraft and solved the range problem of the early turbojet airplane. Commercial long-range aircraft did not use this technique, however, due to safety concerns regarding two large airplanes flying within 20 to 30 feet of one another for extended periods of time, the weight of fuel piping manifolds in the receiver aircraft necessary for aerial refueling, and the cost of maintaining a fleet of tanker aircraft to allow regular commercial service.

The Boeing Model 377 Stratocruiser was a pressurized double-deck luxury piston-powered airliner that could carry 85 passengers in comfort above the weather between the United States and Europe or Hawaii. Reminiscent of the giant ocean-spanning flying boats of the 1930s, this aircraft took about 14 hours to cross the Atlantic and 10 to cross the United States. This was the airplane that the first jetliners would have to compete with in establishing jet aircraft as the principal means of transportation over long distances. (Mike Machat Collection)

Transcontinental Supersonic Speed Records

Not only were absolute world speed records now being set by operational jet fighters, but point to point destination records were attaining supersonic speeds in an American intra-service competition reminiscent of the Pulitzer trophy races of the 1920s. The year 1957 marked the 50th anniversary of military aviation in the United States, and both the air force and the navy set out

Boeing's ground-breaking Model 367-80 served as a proof-of-concept testbed to develop operational practices for jet commerce. Only one "Dash 80" prototype was built, but it proved to be wildly successful in verifying the principle of commercial jet travel in the mid-1950s. This aircraft also encouraged the U.S. Air Force to order hundreds of a similar but larger design as a jet tanker to refuel its B-47 and B-52 fleets. Boeing test pilot "Tex" Johnston performed two barrel rolls in the Dash 80 during hydroplane races near Seattle in 1955. When questioned about why he did that by Boeing president Bill Allen, Tex replied calmly, "I was just selling airplanes." (Mike Machat Collection)

to commemorate the occasion with a bang (or "boom" as an American observer called it). The goal was to fly nonstop across America, coast to coast, as fast as possible with the average speed being supersonic. Aerial refueling now made this possible. Although operational jet fighters could reach speeds of over Mach 1.5 at cruise, they then had to slow down and descend from their cruise altitude to refuel from aerial tankers multiple times during the flight.

The navy struck first with Project Bullet, a flight from California to New York using two new Chance Vought F8U-1P Crusaders. One of the Crusaders damaged its refueling probe early in the mission and had to retire, leaving a lone airplane to land at Floyd Bennett Naval Air Station on Long Island after flying 3 hours and 23 minutes. It covered 2,446 miles at an average speed of 724 mph, or Mach 1.1, making it the first supersonic coast-to-coast speed record. The aircraft was a photo reconnaissance version of the Crusader, and as a bonus, took a series of panoramic photos of the entire country during the flight. The highest speed actually achieved was Mach 1.7 and the name of the pilot was Marine Maj. John H. Glenn, Jr., a combat veteran who flew the Sabre jet in Korea and scored

Pan American World Airways inaugurated jet passenger service from New York to Paris on 24 October 1958 using this Boeing 707-120. Mounting engines on pylons under the wing allowed larger upgraded powerplants to be installed more easily without having to completely redesign the airplane. Thus, although the British Comet was a masterpiece of streamlining with its engines buried within the wing, the Boeing design with externally podded engines set the standard for most subsonic jet airliners, even today. (Mike Machat Collection)

The new, larger, and more powerful Boeing 707-320 Intercontinental, shown here in the markings of Trans World Airlines, became the world's first true long-range jetliner able to link every continent on the globe while carrying up to 150 passengers. Although initial versions of the -320 were powered by turbojets, the improved turbofan engines seen here ensured the Intercontinental's success until it was superseded by Boeing's own 747. (Mike Machat Collection)

America's second jetliner to fly was the Douglas DC-8, built in Long Beach, California. Initially offering longer range than the 707, more DC-8s were ordered by Pan American than 707s in 1955, but the Boeing jet soon overtook orders for the Douglas transport. The DC-8 has an unusual distinction, however, of setting an altitude record of 51,000 feet during flight testing, and becoming the world's first supersonic airliner (by technicality) when a DC-8 Series-40 achieved a calibrated airspeed of Mach 1.05 in a power dive over Edwards in 1961. (Mike Machat Collection)

several MiG-15 kills. He became one of the original Project Mercury astronauts and was the first American to orbit the Earth in 1962. Glenn was subsequently elected as a U.S. senator and ironically played a part in ensuring that an air force reconnaissance airplane would shatter his own record in 1990.

Not to be eclipsed by the navy, the air force proceeded with Operation Sun Run on 27 November 1957, using a small fleet of photo-recon McDonnell RF-101 Voodoos and new Boeing KC-135 jet tankers. The KC-135s allowed the Voodoos to refuel at Mach 0.8 (rather than the less-than-300-mph speed the Crusaders had to maintain to refuel), and they set new coast-to-coast records of just over 3 hours in each direction, demonstrating that air force fighters could now cross the country in both directions in less than 7 hours using inflight refueling from jet tankers.

Back in 1925, the United States Congress had ended military competition at the National Air Races because they considered the effort to be too expensive. By comparison, this modern navy/air force air power competition and speed demonstration probably far outstripped the cost of all the Pulitzer trophy competitions combined, but as a celebration of 50 years of military airpower com-

Convair's entry into the jet sweepstakes was the sleek Model 880 (for its 880 feet-per-second cruise speed), and was the most esthetically beautiful first-generation jet airliner. Seating five abreast instead of the 707 and DC-8's six across, the 880 served short- and medium-range routes for its worldwide airline customers. Convair's choice of General Electric CJ-805 turbojets created problems in commonality for maintenance and training, however, and that coupled with excessive fuel consumption led to only 65 of these aircraft being built. (Mike Machat Collection)

Britain's elegant 80-passenger Comet 4 was an improved development of the ill-fated 44-passenger Comet 1, which first flew in 1952. This British Overseas Airways Corporation Comet 4 inaugurated the world's first jet passenger service across the Atlantic on 4 October 1958 by flying airline passengers from London to New York in only half the time it took in a piston-powered Douglas DC-7, Boeing Stratocruiser, or Lockheed Constellation. (Mike Machat Collection)

This is what America's first truly supersonic-capable engine looked like: the incomparable Pratt & Whitney J57. Powering every new U.S jet aircraft from the North American F-100 Super Sabre and secret Lockheed U-2 photo-recon spy plane to the Boeing 707 and Douglas DC-8 jetliners, the J57 produced up to 14,500 pounds of thrust in afterburner. (Compare this scene with the photo of the first GE 1-A turbojet on page 33.)

bined with the fact that this all occurred within three months after the Soviet Union launched Sputnik, no one really complained about having favorable publicity.

Commercial Jet Travel Replaces Propellers and Ocean Liners

Hard lessons learned in the disastrous introduction of the British Comet 1 jetliner took some time to digest. The jet engine coupled with pressurized cabins meant that passenger-carrying airplanes could fly faster and higher than ever before with the main limitation being range. The loss of so many Comets within two years made the industry wonder if there was a basic flaw in jet propulsion placing the entire commercial venture at risk.

After World War II, Boeing converted the technology of their long-range pressurized B-29 to produce the Boeing Stratocruiser, a more powerful double-deck version that could carry passengers in luxury above the weather between the U.S. and Europe or Hawaii. The number of long hard-surface runways in the world had increased tremendously during World War II due to military requirements for carrying cargo to far-off locations. Now those same runways could be used by civilian passenger aircraft to carry large numbers of people to far-off destinations. The question was, would those passengers be carried by piston-powered airliners or jet-powered aircraft?

In the early 1950s, Boeing produced more than 2,000 swept-wing B-47 Stratojets and later a total of 700 intercontinental-range B-52 Stratofortresses. In order to keep secret the fact that they were building a better airliner, Boeing deliberately gave it the misleading model number "367–80." This pioneering jet transport prototype served as a proof-of-concept testbed to develop operational practices for jet commerce. The single "Dash 80" proved to be wildly successful as it encouraged the U.S. Air Force to order hundreds of a similar design to refuel the B-47 and B-52 fleets, as well as jet fighters. With Boeing

Commercial Jet Airliners: The Russians Race to Be First

While the British and Americans were developing jetliners from scratch, the Soviet Union was taking a different approach to the jet airliner race. Aircraft manufacturer Andrei Tupolev built medium- and long-range jet bombers for the Soviet Air Force. To win the prestige of having jet airliners, the Soviet Union had Tupolev make passenger versions of both the twin-engine Tu-16 medium bomber (known to NATO as "Badger") and the four-engine turboprop Tu-95 long-range bomber ("Bear").

The passenger-carrying versions of these two aircraft received the NATO code names of "Camel" and "Cleat." The Tupolev Tu-104 (designation of the twin-jet airliner) had engines initially of 19,000 pounds thrust each and could carry 50 passengers approximately 2,000 miles. In 1956, three of these evolutionary aircraft appeared in London bringing Soviet government officials and their entourage for a conference, completely taking the aviation world by storm. The resemblance to the Badger bomber was obvious, as these airliners had even retained the military bombardier-type glass nose of the early Badgers.

In 1957 the Soviet Union used the Tu-104 to set world speed records of 971 kmph (583 mph) while carrying large payloads up to 10 tons. The date was 24 September 1957, and less than two weeks later the Soviet Union surprised the world even more by launching the world's first artificial Earth satellite, Sputnik 1. The Soviet aviation industry was not finished with surprises, however.

A Soviet delegation arrived in New York City in 1958 carried by the world's largest transport, the Tupolev Tu-114, the first true intercontinental "jet-powered" airliner. It had a 35-degree swept wing and four turboprop engines of 12,000 pounds shaft horsepower each, and each with twin contra-rotating four-blade propellers with near-supersonic tip speeds.

The Bear bomber from which it was adapted had been intended as a backup for the Myasishchev Bison jet bomber, powered by four large turbojet engines. Both aircraft were intended for strategic delivery of large and bulky Russian atomic bombs over intercontinental ranges. The Bison, however, was unable to achieve the desired range and only 40 were built. Most were converted to jet tankers. Ironically, the swept-wing Tupolev turboprop was almost as fast as the jet bomber, and several hundred were built. Many of these iconic veteran Russian bombers are still soldiering-on in long-range service today.

The arrival of this mammoth passenger-carrying aircraft in New York on a nonstop flight from Moscow was as much of a surprise as the medium-range Tu-104's arrival in London two years earlier. This came just as the United States and Britain were locked in the race to begin transatlantic jet service. The Tu-114 not only had longer range than either the improved Comet 4 or Boeing 707, but could also carry 116 passengers' more than 5,000 miles at 478 mph, beating propliners like the DC-7C and Super-G Constellation by more

Tupolev Tu-104 medium-range jet transport at a regional Russian airport. The Tu-104 was somewhat retro when still being used in the early 1970s, but was nevertheless a comfortable aircraft for the passenger. Note the original "bombardier nose" alluding to the aircraft's design origins as the Tu-16 Badger bomber. Despite all the progress made in introducing commercial jet transports in Europe and the United States after the ill-fated Comet 1's short reign, the Tu-104 was the first jet airliner to be used in sustained regularly scheduled commercial service. (Wings & Airpower Historical Archive)

Ant's-eye view of the massive Soviet Tupolev Tu-114 turboprop transport taxiing out for takeoff. This aircraft was the world's first true intercontinental turbine-powered airliner; it was used regularly by Aeroflot Russian Airlines to carry 170 passengers between the Soviet Union and Cuba. With its ear-splitting noise and noticeable vibration, an eight-hour flight in the Tu-114 was an unforgettable experience, illustrating why having 32 transonic propeller tips on one airplane is not a good idea for passenger comfort or cruise-propulsive efficiency. (Wings & Airpower Historical Archive)

than 120 mph. In 1962, the FAI awarded the special Paul Tissandier diploma to test pilot Ivan Sookhomlin for establishing more than 30 world speed, altitude, range, and payload records for turboprop aircraft in the Tu-114, many of which still stand today.

The author had the opportunity to fly in both the Tu-104 and Tu-114 while on vacation in the USSR in 1972. The Tupolev 104's passenger cabin, although somewhat dated, was smooth and comfortable and a pleasant experience, if not particularly memorable. (In that era, stewardesses on Aeroflot airliners handed out hard candies before takeoff and prior to descents. Sucking on the candies relieved passenger's ear blockages, as the pressurization system seemed to have difficulty keeping up with altitude changes.) The Tu-114, however, was definitely memorable because the first-class passenger cabin in which the author was seated was directly adja-

cent to the inboard propeller discs of the monster turboprop engines.

The Tu-114's cabin was presumably soundproofed to the maximum extent possible, but the vibration of the nearby propellers was best described as mind shattering. An apple placed on the wooden meal tray vibrated its way across the tray and into the aisle. The eight-hour flight between Moscow and Novosibirsk was described by common agreement of passengers as like riding in the barrel of an operating cement mixer. Modern F-15 pilots who had intercepted the Bear bomber version mentioned that it was the only target they've ever intercepted where you could actually feel the vibration of the propellers in the cockpit of the interceptor within several hundred yards. This Tupolev long-distance turboprop remains a unique airplane in aviation history.

This study of Pratt & Whitney JT3D-3 turbofan engines on a TWA Boeing 707-131B shows the larger fan section cowling forward of the "core section" of the engine. The fan produces a shroud of cooler air surrounding the core nacelle and engine exhaust, providing added thrust, reduced fuel consumption, and lower noise levels. Thrust of the JT3D fanjet was 16,000 pounds without the need for smoke-producing water injection. (Mike Machat)

piston engines. In the 1950s, airline fares were dictated by the Civil Aeronautics Board and the only competition between airlines was in schedule timing and quality of passenger service. The comment was made that once one airline bought jet airliners all the rest would have no choice but to follow suit. This, indeed, proved to be the case as the advertising from individual airlines emphasized nothing but jets, once an airline had made the decision to purchase them.

Even airliners powered by turboprops (a jet engine driving a propeller powered by an additional turbine) used the words "jet power" in their advertising to emphasize that they were using the latest technology. The late 1950s also coincided with an increased interest by the public in jets and rockets, especially emphasized once the space race began with the Soviet launch of Sputnik. As a result, just as there had been a huge wave of public enthusiasm for hot-air balloons in the early days of aviation, enthusiasm for high-speed jet airplanes was so strong that even cars and restaurants sported fins and sleek swept-wing motifs. The jet airliner was definitely the wave of the future.

The swept-wing Boeing 707 first flew in 1954 and was quite different from the 1947 design of the Comet jetliner. The wing was swept back 35 degrees and the engine nacelles were suspended on pylons beneath the wings to minimize damage to the fuselage should a jet engine come apart in flight. Since the B-47 was a tricky airplane to fly due to its thin wing, mainly when maneuvering at high-altitude, the 707 was designed with a somewhat thicker wing and with multiple high-lift devices such as flaps and slats to aid the pilot in landing safely. It also had two sets of ailerons to ensure redundancy in banking the aircraft. One of the biggest differences, however, was in the engines, which dictated the overall size of the airplane.

The Comet was driven by four Ghost engines of 5,000 pounds of static thrust each. The prototype 367–80 had four engines of 10,000 pounds thrust each and the operational 707 aircraft had four turbojets producing more than 12,000 pounds thrust each. Hanging

having this contract in hand, the airlines had much more confidence in buying fleets of similar jet airliners to carry passengers.

All those who flew on the Dash 80 became instant converts to jet airliner travel. Propliners such as the Lockheed Super Constellation and Douglas DC-7 took approximately 10 hours to fly coast to coast in the United States and 13 to 14 hours from the United States to Europe at 300 to 350 mph. The jets, on the other hand, cut that travel time almost in half.

The other advantage of turbojet aircraft was their smoothness of ride versus the propeller noise and vibration of huge, powerful the engines from the wing allowed larger upgraded engines to be installed more easily without having to completely redesign the wing. More powerful engines meant larger payloads, and a stretched fuselage to accommodate more seats. As a result of this configuration difference, the first Comet jetliner could seat only 44 passengers whereas the later intercontinental-range Boeing 707-320 could carry as many as 180 passengers across the Atlantic without refueling. Thus, although the Comet was a masterpiece of streamlining with the engines buried within the wing, the Boeing design with external engine nacelles set the standard for most subsonic jet airliners even to this day.

Jet Airliners Arrive in America

Although the Boeing Company had stolen a march on its competitors by building the Dash-80 with their own financing, it was

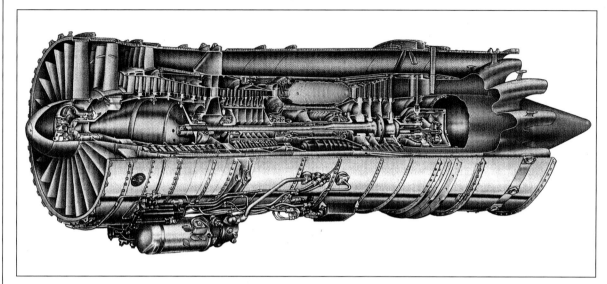

Cutaway of the Pratt & Whitney JT8D-209 medium bypass-ratio fanjet used on the McDonnell Douglas DC-9 Super 80 (MD80), which first flew in 1979. The Super 80 was a record breaker of sorts, having the lowest noise levels, fuel consumptions, and emissions of any commercial jet aircraft upon its introduction to airline service. Certain models of privately owned Boeing 727s (first flown in 1963) have been retro-fitted with the JT8D-209 to meet modern-day noise restrictions.

This MiG-15 acquired by the U.S. Air Force is under guard at an American air base. Notice the Sabre jet lineup in the background. The MiG-15 used copies of British jet engines, but the aircraft was smaller and lighter than the Sabre jet and hence was able to out-climb the Sabre as well as have a several-thousand-foot advantage in cruise ceiling due to its lower wing loading. (USAF)

F-104 Starfighter, known as "the missile with a man in it," is seen here with Lockheed test pilot Herman "Fish" Salmon posing in the cockpit. On a trip to Korea Kelly Johnson heard American fighter pilots complaining about the ceiling and climb-rate advantage of the MiG-15 and decided that a next-generation air superiority machine beyond the F-100 Super Sabre should be lightweight and with a high thrust-to-weight ratio, allowing a higher ceiling and climb rate. (Wings & Airpower Historical Archive)

Use of the Whitcomb area rule was the solution to the underwhelming performance of Convair's barrel-fuselage YF-102. In essence, the area rule redesign redefined the cross-section area of a generic delta-wing configuration, thus reducing drag in the transonic speed regime. These configuration changes were employed on the YF-102A and production F-102A, which became the air force's first supersonic jet interceptor.

obvious that jet airliners were now the wave of the future, and any airline that did not follow suit would be trampled into the dustbin of history. Douglas Aircraft Company, the other major manufacturer of airliners in the United States, rushed to build its DC-8 airliner of similar configuration to the Boeing 707, but with a more conservative, 30-degree swept wing. Convair followed behind Douglas with an airplane again of similar configuration, but slightly smaller, known initially as the Convair 600. It was hoped the jet would cruise at 600 mph, but when that wasn't possible, it was renamed the Convair 880 for its 880-feet-per-second cruise speed.

Boeing and Douglas both chose the Pratt & Whitney JT3 turbojet, while Convair went with a non-afterburning civilian version of the General Electric J 79 engine used in their Mach 2 B-58 Hustler. So by 1960, three different types of jet airliners were being produced by the American aircraft industry, all of them looking like a Boeing 707. Their jet engines all produced approximately 12,000 pounds of thrust and these aircraft were being designed to carry between 100 and 180 passengers coast to coast or across the Atlantic to Europe without stopping to refuel.

Although the race to initiate jet service was actually won by the Russians with the Tu-104 entering European midrange service in 1956, the prestige prize was to initiate transatlantic jet service from the United States to Europe. In addition to the jets' technical and operational issues, there remained one major challenge to meet that goal: noise. The New York Port Authority controlled Idlewild and LaGuardia Airports, with the former being the destination for jet airliners coming from Europe. The noise footprints for these aircraft were considerably different from propliners, being louder for a longer period of time. Local neighborhood groups on Long Island had complained vigorously about the noise of the propeller-driven airplanes and they were even more incensed when they heard early Boeing 707 and DC-8 aircraft operating into New York on trial flights.

Considerable pressure was put on the Port Authority to restrict jet operations, especially at night. The de Havilland Comet 4, which competed with the American airliners, had its engines buried in the wing, thus reducing their turbine whine. For a while it looked as if the 707 would not be allowed to fly into New York because flight paths were over heavily built-up areas. The target date was October 1958 for transatlantic service and it was a race to determine if the 707 would be allowed to land in New York, whereas the quieter British aircraft would. Transatlantic service began 4 October with British Overseas Airways Corporation (BOAC) winning the race by two weeks before Pan Am started its 707 service. But help was on the way with a new type of jet engine known as the turbofan that not only increased thrust, but also led to quieter takeoffs and landings.

Turbofan: More Thrust and Less Noise = More Fuel Efficiency

By the late 1950s most turbojet engines used multiple axial-flow compressors rather than the centrifugal flow compressors of the early turbojets. The twin-spool design that Pratt & Whitney and eventually General Electric used in their engines allowed one turbine to drive the low-pressure compressors while another turbine spinning at a different rate powered the high-pressure compressors. This led to increased efficiency, as was noted in the J57 military engine, which Pratt & Whitney designated the JT3C for civilian usage and which produced about 12,000 pounds of thrust. What was needed was a way to increase the thrust without increasing the size of the engine or the noise it produced, and Pratt & Whitney came up with the idea of converting the J57 into a forward-fan engine to provide added thrust. This new engine literally revolutionized commercial aviation.

The first three stages of the low-speed compressor were removed, and in their place a large two-stage fan with blades that were wider than the engine itself were installed at the front of the engine. The air produced by these larger compressors bypassed the combustion chambers and was routed along the outside of the engine. This air was slower moving than the full compression and combustion chamber exhaust, but on the other hand it was cooler and still produced thrust by moving a larger quantity of air, albeit at a lower velocity than the main turbojet. Thrust was increased by an appreciable 20 to 30 percent depending upon how much air was bypassed through the fan. The larger the bypass ratio, the greater the increase in thrust.

A further benefit was that the slower-moving cooler air mixed with and actually reduced the noise produced by the hot high-velocity turbojet exhaust. The thrust of the JT3 rose from 12,000 to 17,000 pounds, which was equivalent to adding a fifth engine to the Boeing 707 and DC-8 at minimal cost. The fan was driven by a third-stage turbine, which allowed the fan to spin at an optimal rate for producing thrust, leading to further improvement in specific fuel consumption. As a result, within two to three years virtually all American jet airliners had converted to turbofans. Even the New York Port Authority had to admit that noise levels over Long Island were more bearable. Not only were these engines quieter, but their additional thrust allowed jetliners to climb faster, reducing their noise impact more quickly with distance.

In a sense, the turbofan was an optimal marriage between the turbojet and the original classic propeller, with benefits that allowed commercial jetliners to become the major long-distance global mode of transportation. A somewhat odd byproduct of the success of the first Boeing 707s with turbofans was that fewer FAI speed records were being set by jet airliners because they all flew at approximately the same speed, with Mach-1 capability being the physical limiter. There was generally only a 20- or 30-mph difference between top speed and best cruise speeds for jet airliners, and generally only a 20- or 30-mph difference between the fastest (Convair's 880 and 990) and the more stately Boeing 707 and Douglas DC-8.

With the turbofan it became obvious that the method to increase engine thrust was not with an afterburner, but with a larger fan producing a larger bypass ratio between the fan's cool airstream and the core jet engine's hot exhaust stream. For supersonic jet fighters, it was not possible to fit large fans with high bypass ratios into their fuselages, and afterburners remained important for aerial combat. As a result, large aircraft were outfitted with jet engines with larger and larger fans with bypass ratios of 10:1 while supersonic jet fighters had engines with bypass ratios of less than 2:1. Pratt & Whitney's afterburning F100 turbofan installed in the F-15 and F-16 is a good example. But the increased thrust and efficiency of the turbofan allowed large aircraft that didn't require high speed to have greater thrust with fewer engines, leading to vastly increased fuel economy.

The revolutionary Boeing 747 wide-body had four engines configured like the 707's, each one of which produced on the order of 50,000 pounds of thrust. For short- to medium-range flights Boeing developed a twin-engine 737 with engines mounted below the wing on very short pylons. As the need for thrust increased the fans became larger and Boeing's designers had to flatten the bottom of the inlet to allow for the larger fan section. In the 21st Century we have twin-jet aircraft carrying almost the same number of passengers as the 747 with each individual engine producing up to 100,000 pounds of thrust. The era of the large four-engine long-range airliner is finally over.

Mach 2 in the Cold War: A Tale of Two Fighters

The Korean conflict was unexpected but turned out to be a proving ground for the new swept-wing jet fighters with speeds approaching supersonic. The American F-86 Sabre jet was pitted against the Russian MiG-15 and initially claimed a 13:1 kill ratio in favor of the Sabre, but after the war that ratio was downgraded to 8:1. The F-100 Super Sabre capable of supersonic speeds while in level flight was on the drawing boards, but the U.S. Air Force was looking at what the jet fighter beyond the F-100 should be. Lockheed Chief Designer Kelly Johnson went to Korea and talked to Sabre pilots, who all complained bitterly that the MiG-15 could out-climb the Sabre and its ceiling was several thousand feet higher. The air force wanted an air superiority machine that could reach Mach 2, and the take-away Johnson had from his conversations in Korea was that fighter pilots wanted an airplane that could climb fast and reach high altitudes, presumably while also capable of Mach 2 flight.

Johnson was aware that beyond Mach 1.2, wing sweep didn't matter as much as a low thickness-to-chord ratio. He was aware of the Douglas X-3's wing design, which was more a trapezoidal shape. Based upon ample thrust of the soon-to-be-available General Electric J79 turbojet, he decided that would be the optimal wing shape for the air superiority machine known as the F-104 Starfighter. With a

staggeringly low wing thickness-to-chord ratio of 3.36, to save structural weight and to keep the horizontal stabilizer out of the wing wake, the Starfighter was equipped with a T-tail that presented problems for an ejection seat clearing it, thus necessitating a downward ejection seat, similar to the Douglas X-3. This feature was universally disliked by the pilots and led to several test pilot deaths because early J79 engines were prone to failure shortly after takeoff.

Eventually the F-104 was provided with a more conventional upward ejection seat using a rocket motor rather than a pure explosive charge to clear the tail. The F-104 was initially to be equipped with twin 30mm cannons, but that was changed to a single 20mm Vulcan Rotary cannon firing 6,000 rounds a minute, a weapon that survives to this day. The air force specified arming the Starfighter with two Sidewinder infrared-guided air-to-air missiles mounted on the wingtips, as the short span of the wing did not allow for carrying much ordnance on pylons, especially at Mach 2.

Convair's delta-wing XF-92 was good and the F-102 was even better, but they saved the best for last with their brilliantly designed F-106, considered the best dedicated interceptor aircraft ever flown. Clearly visible here is the Dart's area rule fuselage that reduces the formation of drag-inducing shockwaves on the aircraft's elegant delta wing. (Wings & Airpower Historical Archive)

This ant's-eye view of the F-105B shows how deceptive a two-dimensional three-view drawing can look when compared to the real thing. Note the dynamic look of the Thunderchief's area-rule fuselage at the wing root, as well as the twin "sugar scoop" air intakes and low-set stabilators. (Mike Machat Collection)

The prototype XF-104 first flew in March 1954 at Edwards AFB. The GE J79 jet engine was not yet ready, so a Wright J65 afterburning engine with 2,000 pounds less thrust was substituted. The 104's extremely thin wings had sharp leading edges requiring felt guard strips to be installed, allegedly so ground crews did not cut themselves. The wings were also tiny, with a span less than 22 feet and a wing area of only 179 feet, leading to a wing loading for takeoff of an inordinately high 88 pounds per square foot for the prototype, and an even worse 115 pounds per square foot for the operational F-104A. The popular press referred to the F-104 as "a missile with a man in it," and numerous photos were published of a pilot in a partial pressure suit peering out of the cockpit looking like a creature from outer space.

The truth was somewhat different. In its first seven years of service the U.S. Air Force lost 49 Starfighters with 18 pilots. It was found that with its high wing loading the 104 was not a good dogfighting airplane. In December 1958, the F-104 order was cut from 722 to 155, and the aircraft was turned over to the air national guard. While the operational aircraft were proving to be somewhat of a disappointment for their original mission, a YF-104A equipped with the J79 engine based at Edwards took both the world speed and absolute altitude records in May 1958, the first time in FAI history that one type of aircraft held both records. The altitude was 91,250 feet and the speed record was 1,404 mph (Mach 2.13).

Meanwhile, on the other side of the Iron Curtain, the Soviet Air Force was looking for a fighter airplane to do essentially what the

REPUBLIC F-105 THUNDERCHIEF

This factory three-view silhouette drawing of the Republic F-105B shows how the jet almost looks like three different aircraft depending on the view. (Republic Aviation Corporation)

Visually defining the term Jet Age, Grumman's supersonic F11F-1F Tiger shows off its "pinched" wasp-waist area-rule fuselage designed to reduce transonic drag. Also known as a "Coke bottle" fuselage, this design helped the Tiger achieve high performance, including an FAI world record altitude flight to 76,932 ft. in a zoom-climb over Edwards on 18 April 1958. (Wings & Airpower Historical Archive)

F-104 was supposed to do: go to Mach 2 and climb rapidly. Initially it was intended to intercept high-altitude enemy bombers carrying nuclear weapons but it was also expected to be able to fight opposing jet fighters, as was the case in Korea. The MiG design bureau was chosen to explore the possibilities and they built several prototype aircraft, one with highly swept wings and the other with the delta wing, but one also added a conventional horizontal tail. The fuselage of both fighters was essentially identical and the only visible difference was in the type of wing. The swept-wing airplane flew first in February of 1955 while the tailed Delta airplanes flew during 1955 and 1956.

The prototypes received the NATO identifying names of Faceplate and Fishbed. Confusion ensued in the West as to which airplane had actually gone in service with the vote going to the highly swept-wing Faceplate. In reality, the Fishbed was the winner and it became known as the MiG-21. The engine produced only 13,000

This portrait of Richard Whitcomb shows the aeronautical engineer with several of his notable designs, including the area-rule fuselage and super-critical wing airfoil shape. (Craig Kodera)

pounds of thrust in afterburner, but to counteract that, the MiG-21's empty weight was 2,000 pounds lighter than the F-104. But with a delta wing, its wing loading was a more reasonable 66 pounds per square foot, leading to better takeoff and landing performance than the 104, as well as more agility in dogfights at high altitude.

The delta-wing configuration was a model of simplicity with the thickness-to-chord ratio of approximately 4½ percent, and the 57-degree swept delta had a single, short fence in front of the aileron near the wing tip to maximize control effectiveness. The all-moving horizontal stabilizer used to control pitch was mounted low so that at high angles of attack the disturbed airflow did not blank out the tail.

The main complaints about the MiG-21 were its limited fuel and armament. Like the initial six F-104s, it was originally planned to have two 30mm guns, but one of the guns was removed to allow installation of two infrared guided missiles. (By coincidence, these were Russian copies of stolen American Sidewinder missiles similar to the F-104's.) Eventually, even the one cannon was removed. The lightweight MiG-21 indeed proved to be a match for the F-104 when on 31 October 1959 a MiG-21 designated for record purposes as the "YE-166" took the absolute speed record back from the Americans at 1,482 mph (Mach 2.25). In 1961 it set a new absolute altitude record of 113,862 feet, although this was done with the assistance of a 6,600-pound-thrust rocket motor in addition to the afterburning turbojet.

It's interesting to note that despite both the Starfighter and MiG-21 being designed to solve the same problems of Mach 2 fighter aircraft, in reality, both aircraft seldom ever flew at that speed. The thrust of their engines was used instead for climbing, maneuvering, and for flying at high subsonic speeds. Both aircraft were modified to allow flying at subsonic speeds at low level while carrying bombs. The F-104G Starfighter was produced in large numbers for NATO countries, as well as for the Japanese and German air arms. The F-104G's empty weight was 3,000 pounds greater than the F-104A, but it had longer range and an improved J79 for low-level penetration of Soviet air defenses while carrying a 2,000-pound nuclear weapon.

The MiG-21 faced its first test of fire against Western aircraft in the Mideast Six-Day War in June 1967. Its principal opponent was the French-built Mirage 3C flown by Israeli pilots, and in dogfight competitions at medium to low altitude the French-built airplane proved to be superior. This led to the realization that the MiG-21 was under-armed, and additional underwing pylons were added to allow more missiles to be carried, as well as bombs, similar to the Starfighter experience. A 23mm rapid-fire cannon was also installed to supplement the air-to-air missiles.

A new engine was later installed with an additional 1,000 pounds of thrust in afterburner, as well as improved avionics and more fuel in the form of a rather ugly saddle tank on the spine of the aircraft, which added more fuel with very little additional drag. This later-model Fishbed was used until the closing stages of the Vietnam War, where the MiG-21 had held its own against American air superiority machines such as the Phantom and the Crusader. The heavily outnumbered North Vietnamese Air Force had few MiG-21s, but the ones they had were used effectively to interfere with the American's bombing campaign against North Vietnam.

In terms of kill ratios, American fighter pilots were no longer as well-trained as they'd been in the Korean War. In overall numbers, the American victory ratio was only 3:1, and sometimes only at a breakeven point. But the purpose of an air force is to conduct military operations to carry out national policy, not to keep individual pilot scores like a tennis match. Although aerial duels in the late 1960s were between Mach 2 aircraft, none of this combat ever took place at Mach 2. In that sense, the quest for speed was somewhat misdirected because aerial combat was not taking place at multiple supersonic speeds. Instead, increased engine power was used to increase the effectiveness of jets, which spent most of their time subsonic. Nevertheless, the quest for higher speeds continued.

The question will always be raised, which was the better airplane, the F-104 or the MiG-21? Both designs were unanimously praised by their pilots for flying qualities and were described as a joy to fly. But there was always a "but" for both airplanes. Both were considered to be short on fuel and, therefore, flight endurance. The F-104A was essentially rejected by the U.S. Air Force for its primary mission, and less than 300 were built. The other 2,300 or so were bought by American allies and were mainly used in the air-to-ground role (although with improved radars and missiles, the "ultimate" F-104S used by the Italian air force in the 1980s came closest to meeting the goals of the original F-104 design).

MiG-21 production totaled more than 11,000, and at one time, it was described as the most successful supersonic fighter ever built. It was inexpensive, easy to fly, and landed at reasonable speeds due to its low wing loading. It was also simple and easy to maintain, unlike the maintenance problems experienced by the German air force in the 1960s when their Starfighters gained the nickname "widow maker" and led to the fall of the German government due to scandals in the procurement of the airplane.

The ultimate "proof of the pudding" for fighters is a head-to-head aerial battle using real weapons and with equally skilled pilots. This happened in 1971 in the Indian-Pakistani wars when the Pakistanis were flying American F-104C Starfighters and the Indian Air Force was using the Fishbed D (MiG-21 FL). Both air forces had skilled and professional pilots. They met in aerial combat and the score was MiG-21: 4; Starfighter: 0. No combats were flown at Mach 2 and all the engagements were made at low-to-medium altitudes. Modern aerial combat is won sometimes just by the luck of the initial position and who-saw-who first. A better comparison of the merits of these two advanced airplanes might be that most Starfighters today are in museums, whereas the Romanian Air Force only retired their last Fishbeds in 2016, and the National Test Pilot School in Mojave, California, added a vintage MiG-21 to their test fleet that same year.

Area Rule, Supercritical Wings and Winglets

Richard Whitcomb, an aerodynamicist working for NACA and then NASA, had probably more influence on jet age aviation that any other single individual. Described by colleagues as being fairly shy and conservative, when it came to advancing theories and defending them, he was more like a tiger. One person once said that Whitcomb in his brain could see air moving around an airplane better than anyone else could see in a wind tunnel. He used to say part of his philosophy was to set aside part of the day for just sitting at his desk and thinking. That thinking paid off in a big way when

The winglet on a Bombardier Challenger 600 executive jet is designed to modify wingtip vortices to provide lift, reduce drag (corresponding to thrust), and increase the lift/drag of an existing wing with minimal changes. Test results showed a 2-percent improvement in specific fuel consumption compared to a standard non-winglet wing due to drag reduction. (Mike Machat)

he received both the Collier trophy and the National Aero Club award for two completely different discoveries: "Area Rule" and the supercritical wing.

Once the sound barrier had been broken using rockets, the air force decided that future jet fighters should be able to reach supersonic speeds in level flight. Using afterburners to produce extra thrust for a short period of time was a brute force approach, but studies soon showed that although turbojets with afterburners did not run out of gas as soon as a rocket engine, it was a very inefficient method of going supersonic. The problem seemed to be that up to a flight Mach number of 1.2 there is so much drag in level flight that the airplane could not reach Mach 1.2 without diving. In 1949 NACA developed a slotted transonic wind tunnel that allowed collecting model data in that speed regime, and the tunnel gave engineers the ability to study the shockwaves over the models.

Whitcomb, one of the aerodynamicists conducting experiments with the transonic wind tunnel, discovered by Schlieren photography that there was not just a bow shock in front of the airplane's nose but additional shockwaves seemed to occur where the wing began and ended on the fuselage. Originally aerodynamicists concentrated on fuselage shape for supersonic airplanes, dealing separately with wing shockwave drag. Whitcomb realized that for transonic flight, wing and fuselage had to be considered in totality. He theorized that the extra drag from these additional shockwaves where the wing met the fuselage and the rapid cross-sectional area change of the total shape where the wing started was causing an early shockwave in the supersonic flow. He believed that to minimize the shockwave formation it would be better if there was a smooth progression from the nose to the tail of the total cross-sectional area of the body.

This implied that the fuselage should be pinched in where the wing was located. He created wind tunnel models to demonstrate the effect of the "waisting" and confirmed less drag at transonic speed. He presented his theory in a 20-minute presentation at a symposium at NACA Langley in front of Alfred Busemann, who had presented the paper on swept wings in 1935 at the Rome conference, and waited for his reaction. Busemann stood up and turned to the audience and said, "Some people think up half-baked ideas and call it a theory; Whitcomb thinks up a brilliant idea and calls it a rule of thumb." Newspapermen dubbed the area rule the "Coke bottle" theory and even "wasp waist" design.

Convair's YF-102A was supposed to fly Mach 1.4 with its J57 engine, but it stubbornly refused to even go supersonic in a dive. Whitcomb suggested to Convair engineers "pinching" the fuselage where wingspan was greatest and adding contours to the aft fuselage to smooth the cross-sectional area progression. Once that was accomplished, the YF-102A easily went supersonic in level flight. Grumman Aircraft Company also heard of Whitcomb's discovery. They were producing a supersonic single-engine jet fighter for the navy called the F11F-1 Tiger. The wasp waist of the F11F-1 was very obvious and was seen by millions when the jet was flown by the navy's Blue Angels flight demonstration team. For the area rule Richard Whitcomb received his first Colliers trophy.

But Whitcomb was not through. He realized that the conventional NACA subsonic airfoil in use since the 1920s produced a strong normal shockwave well forward of the trailing edge of the wing. This was due to the fact that the airfoil had its peak curvature fairly close to the leading edge of the wing, and as pressure dropped as airflow moved toward the trailing edge, a shockwave formed because of the change in pressure along the top of the wing. This shockwave caused the smooth airflow over the wing to become turbulent, causing separation of the airflow over the aft portion of the wing, thus causing increased drag.

Whitcomb theorized that by having the top surface of a transonic wing be flatter with the peak of the wing being well aft, then no strong normal shockwave would form in the transonic speed regime, thus reducing drag due to shock stall. Wind tunnel tests confirm this, and virtually every airliner wing today has a Whitcomb "supercritical airfoil," as it is known. For this, Richard Whitcomb received the National Aero Club trophy.

In the late 1970s after the fuel crisis following the Middle East October war, emphasis was placed on increasing the efficiency of jet airline travel. This was the era of the jumbo jets where even a 1- to 2-percent reduction in fuel usage per mile flown could lead to savings of tens of thousands of gallons of jet fuel and millions of dollars in cost. Whitcomb thought about the problem and realized that higher lift over drag allowed less fuel burn; he wondered how it would be possible to change a swept wing to increase the lift-to-drag ratio. Traditionally, a high-aspect-ratio wing (large span short cord) is a more efficient wing, but how would it be possible with minimal changes to give all aircraft a higher aspect ratio?

Wing tip vortices exist on real airplanes and he thought it might be possible to modify the wingtips so that the vortices, which normally create only drag, could also provide a small amount of lift and even negative drag (thrust), thus increasing the efficiency of the wing. The air force modified a KC-135 tanker for use as a flying testbed and discovered that by using his "winglet" design, as he called it, it was possible to reduce fuel burn by 1 to 2 percent. As fuel became more plentiful and less expensive after the initial October war shock, less attention was paid to such small increases in efficiency.

Now, in the 21st Century, with renewed emphasis on more efficient airplanes burning less fuel, the winglet concept is back in style on everything from a Boeing 737 to the Airbus A320, as well as almost every executive aircraft flying today. Its new popularity can be thought of as a tribute to a brilliant man who made such a difference to the traveling public, but who was known only to his engineering colleagues. So, when flying in a jet airliner, look out the window at the winglet and lift a toast to the man who could "see" air. In a way, he's responsible for getting you to your destination a little bit earlier and with a cheaper ticket to boot.

In 1961, 56-year-old Jackie Cochran set eight major speed records in a Northrop T-38 Talon, including a 1,000-kilometer closed-course speed of 639.4 mph and sustained altitude of 56,071 feet. Working for Northrop, Cochran and her mentor, Col. Chuck Yeager, test flew the company's T-38 throughout the month of August. Putting the airplane through its paces, Jackie (with Yeager in the back seat) flew chase from 40,000 feet with the X-15. Again with Yeager, she flew Mach 1.3 and reached 51,000 feet on 22 August. Two days later, with Yeager flying chase in an F-100, Cochran shattered the former speed record with an average speed of 844 mph. In October 1961, she flew to 56,800 feet. Although now deceased, Cochran still holds the FAI record for most official FAI records ever set by a single pilot. (The Museum of Flying)

Chapter 6

MACH 3 AND BEYOND: SUPERSONIC CRUISE (1962–1976)

"Triplesonic" was the enticing term created to describe North American's ultra-futuristic XB-70 Valkyrie. Although never committed to mass production, the result of its own complexity and ICBMs becoming America's chief deterrent weapon during the Cold War, the XB-70 was the largest aircraft ever to fly at three times the speed of sound. Seen here is Ship 1 in NASA markings, taking off from Runway 22 at Edwards in 1967. (Wings & Airpower Historical Archive)

aving conquered the demons of the sound barrier, the next aerial frontier became the challenge of conquering the heat barrier, which came to be known by the colloquialism, "Thermal Thicket." Hot enough to literally melt the wings off an aluminum airplane, temperatures caused by the friction of air moving along an aircraft's skin at high supersonic speeds was a formidable challenge for engineers to overcome. Persistence and technology prevailed, and jets were now designed to be capable of flying at Mach 3.

Jet Age Meets Space Age: Breaching the Thermal Thicket

The Cold War between the United States and the Soviet Union continued long after the Korean War had ended in an uneasy truce. Higher altitude and faster jet fighters were produced to ensure that they could shoot down jet bombers capable of delivering nuclear weapons that would destroy a city and its population on a single sortie. But by 1957 a new means of delivering nuclear weapons was being developed called the Intercontinental Ballistic Missile (ICBM), and there was no defense against a warhead arriving from space at 12,000 mph.

Nevertheless, jet bombers and fighters received major attention by air forces in continuing to go faster and higher. In order to top existing speeds of Mach 2 a thermal barrier was being approached that, although better understood than the sound barrier, produced challenges. The stagnation temperature of air molecules at Mach 3 (2,100 mph) exceeds 500 degrees F and aluminum aircraft started to lose structural strength at temperatures above approximately 350

degrees F. New advances in materials had to be made to produce aircraft capable of sustained flight above Mach 2.

Stainless steel was a material known to withstand temperatures of more than 1,000 degrees F, but it was quite heavy for major parts on airplanes, although very strong. A metal known to resist considerably higher temperatures than aluminum was titanium, which was very light and very tough. High Mach flight implied flying at high altitudes above the stratosphere (in the troposphere above 60,000 feet) where the thinness of the air reduced profile drag on the high-speed aircraft.

As early as 1954 military forces and NACA were jointly planning rocket research airplanes capable of hypersonic speeds of Mach 6 and perhaps higher. Designated the X-15, this hypersonic research rocket plane, like the X-1 and X-2 before it, would be carried to altitude by another aircraft and released. The new launch aircraft was a Boeing B-52. An emergency procedure was developed that if the B-52 developed a problem, its pilot would notify the X-15 pilot that he might have to be jettisoned. Since normally the X-15 was carried under the wing like a bomb with no flight controls active, it would be up to the test pilot to try to very quickly start the auxiliary power unit to give himself flight controls so he could attempt to land the X-15 after it was jettisoned.

Also like the X-1, the X-15 was powered by a rocket engine, but one considerably more powerful than previous X-planes that would be capable of being throttled like a normal jet engine. This was necessary to allow a more careful approach to the hypersonic

regime. The four rocket tubes of the X-1's engine, although reliable, had only on/off toggle switches with the pilot controlling thrust by selecting the number of 1,500-pound-thrust tubes to fire. The Bell X-2 had a new, throttleable Curtiss Wright XLR25 2-barrel rocket engine providing a maximum of 15,000 pounds of thrust, but unfortunately throttling was never used on the X-2's 13 powered flights.

The X-15's operating envelope was estimated to be more than Mach 6; with

X-15 Ship 2 being carried to launch altitude by its NB-52A mothership, "Balls Three" (S/N 52-003), is seen venting excess liquid oxygen and anhydrous ammonia propellants through overflow tubes at the aft end of the fuselage. Climb-to-launch altitudes usually took one hour, whereas the X-15's return trip to Edwards was made in about 8 to 10 minutes!

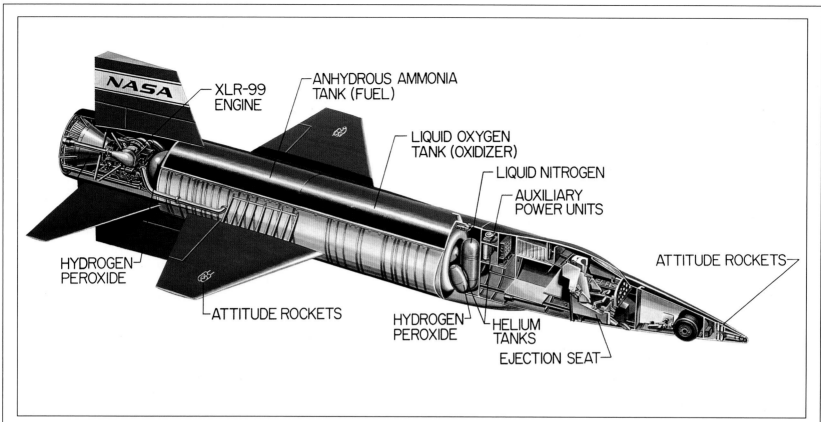

Labels on diagram:
XLR-99 ENGINE
ANHYDROUS AMMONIA TANK (FUEL)
LIQUID OXYGEN TANK (OXIDIZER)
LIQUID NITROGEN
AUXILIARY POWER UNITS
ATTITUDE ROCKETS
HYDROGEN PEROXIDE
ATTITUDE ROCKETS
HYDROGEN PEROXIDE
HELIUM TANKS
EJECTION SEAT

Cutaway view of North American X-15 hypersonic research aircraft. As was customary with rocket research vehicles of this era, this cutaway is dominated by fuel and oxidizer tanks for the rocket engine, which in the X-15's case burned for less than 1½ minutes, yet propelled the airplane to speeds as high as Mach 6. (NASA)

The X-15's massive Reaction Motors XLR99 throttleable rocket engine provided up to 57,000 pounds of vacuum thrust. The engine ran for about 80 seconds before running out of fuel, powering the X-15 to its numerous speed and altitude records before returning to Earth as a powerless high-speed glider. (Author Photo)

altitudes of 250,000 feet, calculations placed the airplane's skin temperature in excess of 1,500 degrees F. As a result, the aircraft was built of Inconel X, a heavy nickel/steel alloy that is strong and can withstand high temperatures. To control the attitude of the aircraft outside the atmosphere, a system of small hydrazine-powered reaction jets would be installed to control pitch, roll, and yaw as the usual aerodynamic control surfaces would be useless in the vacuum of space.

Although the X-15 would never enter orbit, during its descent, the pilot was flying the world's fastest glider, landing on the 3-mile-long Lakebed Runway 18 at Edwards. The rocket plane provided valuable piloting experience for future manned spacecraft that used wings rather than parachutes to land. It is interesting to note that Neil Armstrong, the NASA Astronaut who flew both Gemini and Apollo capsules, also flew the X-15 as a test pilot seven times, and welcomed conversations about that experience even more than his Apollo Lunar mission. He actually regarded flying X-15 profiles as more challenging than flying a spacecraft. The joke regarding manned satellites versus the X-15 was that once the satellite's booster rocket engine quit, Sir Isaac Newton took over. In the X-15, the pilot could still be a hero, or the "goat" if things didn't go so well.

The X-15's joint sponsorship is best depicted in this photo of the original pilots assigned to the program from the U.S. Air Force, U.S. Navy, and NASA. From left are: USN CDR Forrest Peterson; NASA's Neil Armstrong; U.S. Air Force Col. Robert Rushworth, NASA's Jack McKay, U.S. Air Force Major Bob White, and NASA's Chief Rocket Test Pilot, Joe Walker. White flew the X-15 to Mach 6, and Walker made the all-time highest flight reaching 354,000 feet. Rushworth had the most flights of any X-15 pilot. (Wings & Airpower Historical Archive)

USAF Major Robert M. White became the prime air force X-15 pilot when he replaced his best friend and "First of the Spacemen," Capt. Iven C. Kincheloe, who had been assigned to fly the X-15 but was killed in an F-104 accident in July 1958. White flew X-15 Ship 2 (56-6671) to record speeds of Mach 4, Mach 5, and Mach 6, and became the first man to fly higher than 200,000 and then 300,000 feet, all within a time period of only 18 months! (Wings & Airpower Historical Archive)

Studies of the X-15 profile, including altitude, Mach number, and effective wind blast (Q), indicated that an ejection seat and pressure suit would be adequate for pilot escape during most ejection scenarios, eliminating the need for a heavy and complex escape capsule. The X-15's ejection seat was specially built by the Stanley Company, and was extensively tested on high-speed rocket sleds. These tests verified that two extended booms would keep the seat stable under virtually all conditions. (NASA)

Aircraft with high skin temperatures tend to have small windows due to the different thermal expansion of glass versus metal, creating unneeded stresses if the windows are too large. While climbing to launch altitude, the pilot could contemplate the X-15 mission scoreboard visible out his left window. During rocket-powered flight climbing steeply to altitude, the pilot could generally not even see the earth below. (NASA)

Beating the Heat: Hot Structure and Heatsink versus Ablation

Because calculation of hypersonic temperatures in the mid-1950s was still an unknown field, there was a tendency to be extremely conservative in estimating the temperatures and time to reach thermal equilibrium. The question became, "How do you protect the vehicle during the period when those stagnation temperatures at high Mach theoretically reach their highest values?" The X-15 approach was to have a hot structure for the airplane because the hot periods would be extremely brief. Hot structure implied building the vehicle out of materials that could tolerate high temperatures for those short periods of time.

Originally it was estimated the X-15 would be heated to approximately 1,000 degrees F during its experimental sorties with the leading edges of the wings possibly reaching 1,500 degrees F. Subsequently it was estimated that the entire X-15 would probably be heated to 1,500 degrees including the underside of the airplane. In order to account for the increased heat load, the weight of the metal structure of the airplane was increased by about 2,000 pounds to provide a greater heatsink to absorb heat and minimize hotspots during the high-Mach test points, the duration of which generally was only 2 to 4 minutes.

By comparison, the SR-71, which cruised for more than 40 minutes at Mach 3 and higher, reached thermal equilibrium 24 minutes after it reached its cruise speed of Mach 3. Measured skin temperatures varied from 590 degrees F near the nose of the aircraft to 1,100 degrees in the vicinity of the engine nacelles (the increased temperatures at the aft end of the airplane were due to the radiant heat from the afterburner plumes because the aircraft climbed, accelerated, and cruised while in afterburner). That is why aluminum airplanes were not satisfactory for long-term cruise above approximately Mach 2; above 350 degrees F, long-term life of the aluminum structure begins to degrade.

Mach-3 cruise airplanes (of which there have been very few) are generally made of stainless steel or titanium. The X-15 was made of Inconel X, which could withstand extremely high temperatures for relatively short periods of time. The Mach-3 cruise SR-71 was made of titanium, which could resist the stagnation temperatures at Mach 3 indefinitely as the metal was unaffected by sustained temperatures over 800 degrees. The aft fuselage temperatures of 1,000 degrees could be sustained without significant impact for sufficiently long times, since the SR-71 had to slow down to land, or perform an aerial refueling every hour and a half.

The SR-71 fleet flew thousands of hours at Mach 3 and never suffered a thermal structural failure, although the fuselage length of 107 feet actually stretched by nearly 1 foot while at thermal equilibrium temperatures.

Another method for briefly protecting aircraft structure from the heat load of high-speed flight is to provide an ablative coating, which will melt and sublimate during a period of extremely high-heat flight, protecting the structure beneath it by carrying away the heat load. ICBM warheads originally used this method for remaining intact during their relatively brief terminal plunge into the atmosphere at 12,000 mph. Calculating what sort of ablative coating was needed and how thick that coating should be was an art rather than a science in the 1960s, as was proven on the advanced X-15A-2 in its one flight to Mach 6.7.

This graph indicates predictions of temperature profiles as a function of time for a typical X-15 mission based upon structures, metal selection, and thickness. (NASA)

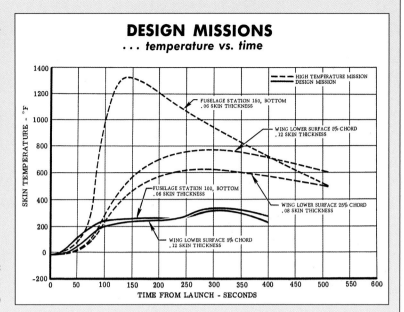

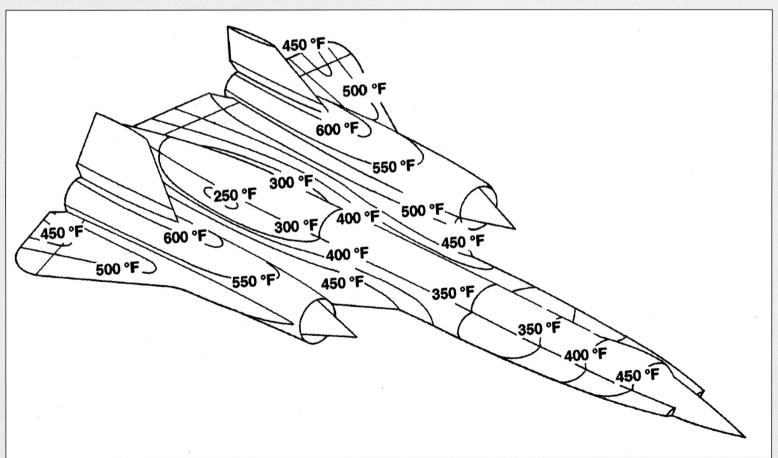

Equilibrium temperature reached on the Lockheed SR-71 Blackbird during sustained Mach-3 cruise. Equilibrium temperatures are typically reached 24 minutes after the aircraft stabilizes in cruise conditions at altitude. (NASA)

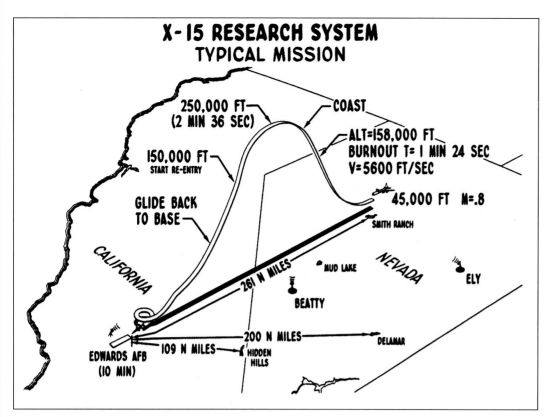

X-15 RESEARCH SYSTEM
TYPICAL MISSION

250,000 FT
(2 MIN 36 SEC)

COAST

ALT=158,000 FT
BURNOUT T= 1 MIN 24 SEC
V=5600 FT/SEC

150,000 FT
START RE-ENTRY

45,000 FT M=.8

GLIDE BACK
TO BASE

SMITH RANCH

CALIFORNIA

NEVADA

261 N MILES

MUD LAKE

ELY

BEATTY

200 N MILES

DELAMAR

109 N MILES

HIDDEN
HILLS

EDWARDS AFB
(10 MIN)

NASA graphic illustrating a typical X-15 mission. High-speed experimental missions were always launched pointing toward the recovery base in the event something went awry. This graphic shows the locations of the dry lakebeds designated as emergency landing points in the event of rocket engine problems. Fire trucks and recovery teams were pre-positioned by support aircraft on these lakebeds the day of each mission.

Ignition of the XLR99 engine after release from the NB-52 mothership at 40,000 feet. Quick ignition was vital to accomplish the pre-planned mission. Failure to ignite required quick action on the part of the pilot to begin jettisoning fuel and oxidizer while preparing to land on a nearby dry lakebed. Successful ignition began an 80-second "'E- ticket'" ride with the pilot pressed back in his seat ("eyeballs-in") at a continually increasing G-force reaching 4 G's by the time the rocket burned out. (NASA)

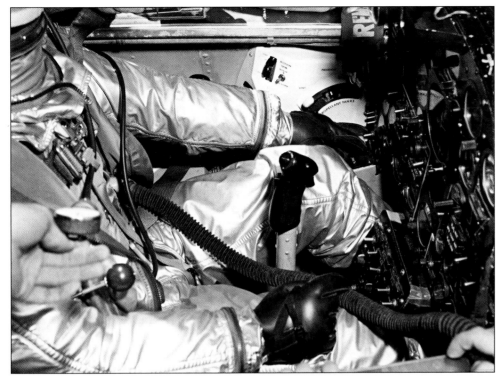

Cockpit interior of the X-15 designed by North American X-15 Project Test Pilot Scott Crossfield. The relatively simple and elegant cockpit had three control sticks with the center-mounted stick and the movable sidestick controller on the right console controlling aerodynamic control surfaces for use within the atmosphere. The left-hand stick on the left console, looking somewhat like an umbrella handle, was used to control the hydrazine thrusters for attitude control above 100,000 feet after the XLR99 had run out of fuel. Most pilots preferred using the right-hand sidestick controller once they became accustomed to it and eventually had the center stick removed to allow more room in the cockpit. (NASA)

X-15 touching down on Rogers Dry Lake with skids deployed. Because there were no brakes on the X-15's landing gear, skid friction on the packed-sand lakebed surface provided the only braking action. The landing gear deployed instantly using gravity for the nose wheel and spring tension for the main skids. In the event the gear did not extend, the pilot had to immediately eject. (Mike Machat)

The throttleable capability of the XLR99 allowed mission planners to develop profiles for the brief X-15–powered flight with much more flexibility in opening the envelope for hypersonic flight. Operating the engine at lower thrust allowed the rocket to burn for a longer period of time, which could be used to trade speed for altitude and vice versa. This led to potentially more complex mission profiles, which in turn led to a more complex process for selecting test objectives for each individual mission. The X-15 was revolutionary for its time in that the mission planners, pilot, and people in the control room during the actual mission could run through the mission profile dozens of times using an analog simulator. Although crude by today's standards, this led to higher success probability and increased productivity on each mission.

Practices included simulated emergencies, requiring the pilot and mission control team to take action to recover the X-15 safely. A typical mission, which only lasted 8 to 10 minutes from release from the B-52, involved some 20 hours of practice by the pilot and engineers before the actual flight was made. The success of this process is illustrated by the fact that two of the three X-15s built are now in museums and only one pilot lost his life during a 199-sortie test program that in truth was extremely dangerous because aircraft and pilot reached speeds never before attained. This mission practice process was copied and followed by all subsequent U.S. orbital manned space flights from Mercury to the Space Shuttle.

Former NACA test pilot Scott Crossfield left that agency to work for North American on the development of the X-15. With his engineering background as well as experience flying the air-launched rocket-powered Douglas Skyrocket, he was a major contributor to the design and ultimate success of the X-15. As the company project test pilot, he was the one who insisted on development of a simulator for practicing mission profiles and emergency procedures for the missions. He essentially designed the cockpit and simplified it based on his knowledge of how quickly the pilot would have to react during rocket burn time. He also knew the stresses a pilot would undergo being smashed back into his seat under acceleration forces of the powerful rocket motor.

It was originally thought that an escape capsule was necessary to protect the pilot in the event he had to leave the airplane in an emergency. Crossfield knew that capsules were dangerously unstable and was convinced that was not the proper escape method. He worked with the David Clark Company to develop a full pressure suit for the pilot for high-altitude protection and also decided that an ejection seat would not only be the lightest and simplest escape method, but also the most stable and least dangerous. The pressure suit coupled with an ejection seat would provide adequate escape capability for the majority of X-15 flight time.

High-speed rocket-sled testing was conducted at Edwards, and it was discovered that two simple folding booms with winglets that deployed into the airstream after ejection would keep the seat stable throughout the entire ejection envelope. The full pressure suit he helped develop served as the forerunner for spacesuits used by Mercury and Gemini astronauts, and later, U-2 and SR-71 flight crews. SR-71 ejection seats coupled with full pressure suits served as the escape system on the first four flights of Space Shuttle *Columbia,* since that was the highest altitude and speed man-rated ejection system in the world.

The X-15 was built with structural limits of plus and minus 12 Gs, although its normal envelope was +7.33 to -2 Gs. Crossfield insisted on having the X-15 built extremely strong based on some wild rides he'd experienced in the Bell X-1. A graphic illustration of the

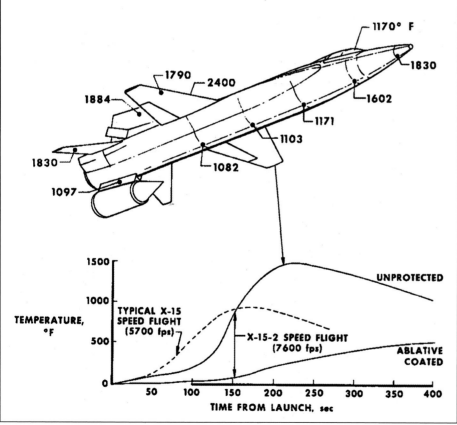

Graphic indicating maximum temperatures predicted for different locations on the X-15A-2 during high-Mach flight. (NASA)

aircraft's robust structure was seen while performing a ground test with the XLR99 engine on Ship 3 at Edwards when the engine exploded with Crossfield in the cockpit wearing normal civilian clothes. The aircraft was literally blown in half with the forward fuselage coming to rest some 30 feet from its original position.

Crossfield said the explosion was indescribable and the G-force was like nothing he'd ever experienced. He remained in the cockpit despite the fire department's attempts to extricate him, as that sealed environment provided the best possible protection against the fire raging around him. An indication of the ultimate strength of the X-15's structure was that, not only was Crossfield completely unscathed during his violent unscheduled rocket ride, but the X-15's aft fuselage was subsequently rebuilt. X-15 number three was put back in service, although that first XLR99 engine was completely destroyed.

X-15 missions began in a rather leisurely fashion. Once the pilot was suited-up and installed into the X-15's cockpit, other than talk to the crew on the intercom, he had nothing to do but look through the rather tiny window and then relax for the 90-minute flight to Utah or Nevada to reach the mission's launch point. The last 12 minutes before launch was a different story, however, as he used an extensive checklist to bring the vehicle to life in preparation for flight, and then reported to the carrier and chase aircraft when he was ready to be released.

The original painting "Launch of the A-2" depicts "Pete" Knight's epic flight of 3 October 1967 at the exact moment of ignition of the X-15's XLR99 rocket engine. The aircraft reached its peak velocity of Mach 6.70 (4,520 mph) after pushing over from 102,000 feet. The launch aircraft seen here was the Boeing NB-52B "Balls Eight" (S/N 52-008), which has been preserved on static display at the North gate of Edwards AFB. The X-15A-2 resides at the National Museum of the Air Force in Dayton, Ohio. (Mike Machat)

X-15 with external tanks and white protective coating over the pink ablative. The amount of ammonia fuel and liquid oxygen needed to extend the burn time of the XLR99 was so high that it was necessary to mount large external fuel tanks with an increased 13,000 pounds of fuel and oxidizer. Although this caused increased drag with lower acceleration, it was hoped that once the external tanks were jettisoned at approximately Mach 2, the original fuel load contained within the X-15's fuselage would allow the aircraft to achieve higher Mach flight. (NASA)

The X-15A-2's sprayed-on pink ablative coating had the consistency of a pencil eraser, and protected the aircraft's Inconel-X skin from the ravages of intense thermo-aerodynamic heating at speeds above Mach 5. A-2 pilot "Pete" Knight loved telling the story of seeing the airplane coated this way for the first time and exclaiming, "There's no way I'm going to fly a @#*%! pink airplane!" (NASA)

X-15A-2 with heat damage after Pete Knight's Mach 6.7 record flight. Overall, the white protective covering over the ablative material looks scorched but most of the damage was on the ventral fin where the white covering and the pink ablative material had melted and holes were burned through the Inconel X metal skin. Hypersonic shockwaves are extremely hot in localized regions and the presence of the scramjet probably caused shockwaves to be generated that caused those hotspots. (NASA)

Convair B-58 Hustler in landing configuration as aircraft 55-665 touches down on Edwards' Runway 04. Note the high angle of attack required to land a delta-wing aircraft. Tires overheating and exploding into the wings were a problem in the early phase of the test program, but that problem was solved when the bomber entered operational service. A total of 116 of these high-performance bombers were built, and 19 world records were set by the B-58 in its relatively short service life. (Wings & Airpower Historical Archive)

Perhaps one of the most distinctive shapes in the sky in the late 1950s was Convair's Mach-2 delta-wing marvel. Evident here is the bomber's pronounced area-rule fuselage, raked delta planform, and four sleek nacelles each housing a 15,000-pound-thrust General Electric J79 turbojet. Top speed of the B-58 was Mach 2.0 (1,360 mph) and maximum range was determined by aerial refueling. It is rare to see a B-58 flying "clean" without its large fuel/weapon pod mounted on the centerline below the fuselage. (Wings & Airpower Historical Archive)

Convair B-58 from below showing the fuel/weapon pod as well as the engine nacelle positions. The 62-foot pod carried a 6,000-pound nuclear warhead and 27,500 pounds of fuel, and was released at the target, leaving a very low-drag airframe to return to base with a considerable fuel saving. While the B-58 was setting speed records from Edwards, the individual Mach 2 legs were only 10 minutes long, but the total flight path covered two states. Official FAI observers were stationed at radar sites on base. (USAF)

B-58 Ship 1 landing at Carswell AFB in Fort Worth, Texas, home of the mile-long Convair manufacturing facility that first built B-24 Liberators during World War II. The B-58's first flight was made here on 11 November 1956, flown by the company's chief test pilot, Beryl A. Erickson. As a result of his extensive flight test work in the B-58, Erickson held the distinction of being the first pilot in the world to log 100 hours of flight time at supersonic speed. (Mike Machat Collection)

Lt. Col. Fitzhugh L. "Fitz" Fulton is all smiles as he and his crew are greeted by Col. Tom Collins after setting the world record for carrying a payload to altitude. The flight was made on 14 September 1962 and the bomber climbed to 85,364 feet carrying a 5,000-kg payload. Maximum speed attained in the acceleration run before pulling up to optimum climb attitude was Mach 2.18 (1,400 mph). Fulton was awarded the Collier Trophy for this achievement, and his record still stands today. (Fitz Fulton Collection)

CIA Lockheed U-2 aircraft at the remote flight test base early in its flight test program. Described as a glider with a jet engine, it was a minimalist design like the F-104, with the exception of its huge 80-foot wingspan. The combination of light weight and high-aspect-ratio wings allowed U-2s to cruise for 8 to 10 hours at an altitude of up to 65,000 feet using a single Pratt & Whitney J57 turbojet without an afterburner. So successful was this design for high altitude and long range that it is still in use, more than 60 years after its first flight. (CIA)

B-58 crew capsule on display at the Edwards Flight Test Museum. Once the crewmember ejected the capsule, he remained in his seat all the way to the ground because he did not have an individual parachute. The capsule was also designed to be airtight to serve as an emergency pressurized capsule that the pilot could use employing rudimentary controls to slow and descend to lower altitudes, meaning the pilot could essentially fly in a shirtsleeve environment without a pressure suit. (Author)

NASA U-2 aircraft family portrait. In the background is a U-2C of the first generation with the 80-foot wing. In the foreground is the modern generation ER-2, which has a larger 102-foot wingspan. The new aircraft are the only ones still flying, but the last NASA U-2C set FAI records in 1998 for a single-engine jet aircraft for sustained horizontal flight at an altitude of 67,190 feet plus time-to-climb, as it was flying to a museum at its home base in the San Francisco Bay Area. (NASA)

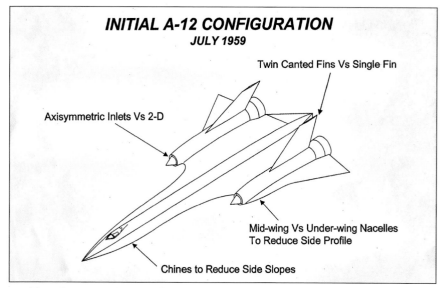

INITIAL A-12 CONFIGURATION
JULY 1959

Twin Canted Fins Vs Single Fin

Axisymmetric Inlets Vs 2-D

Mid-wing Vs Under-wing Nacelles
To Reduce Side Profile

Chines to Reduce Side Slopes

Lockheed A-12 redesign of the A-11 to reduce radar visibility. In an era well before digital computer-aided design systems, the aircraft's redesign was accomplished in approximately two months using only slide rules and small scale models to estimate its radar cross-section. (John Whittenbury)

The B-52 crew then arrived at the correct heading for launch, and once they reached the designated launch point, the X-15 was dropped. Trying to coordinate all this was a rigorous task in itself, and the performance of the B-52 launch platform was critical in ensuring that the mission began correctly. The X-15 had a tendency to always roll to the right on launch, so the pilot had to correct for this because a mid-air collision would ruin the entire day. The next task was to light the rocket engine to start the powered portion of the flight, knowing that in less than 15 minutes, the X-15 would be rolling out on Rogers Dry Lake, some 200 to 300 miles to the south.

NASA test pilot Milt Thompson once said, "[The X-15] is the only airplane I ever flew that I was grateful when the engine quit." Thrust of the XLR99 engine at full throttle was overpowering, especially as the X-15 became lighter with the rocket engine gulping its heavy fuel load. The "eyeballs-in" acceleration rose rapidly due to the steadily increasing thrust-to-weight ratio, and by the time fuel was expended, the pilot was pressed back in his contour-fitting seat at more than four Gs. Because of the X-15's small cockpit windows

The same speed record course flown by Lockheed's YF-12 in 1965 was used for the SR-71 in 1976 to further raise the records. A chart of the routing for the 15-to-25-km straightaway course illustrated that when cruising at Mach 3, simply turning around to reverse direction could encompass up to four states. Much planning went into actually laying out the flight path that the airplane would follow, since the flight time between the timing gates for the 25-km (15 miles) distance took less than 30 seconds!

North American's radical-looking XB-70 rolled out of the Plant 42 manufacturing facility at Palmdale, California, on 11 May 1964. The event was considered to be a watershed moment in aviation history, in that a jet-powered aircraft could fly at a sustained speed of Mach 3 only eight years after the rocket-powered Bell X-2 flew Mach 3 for the first time. Hard to believe this airplane first flew in the same year that the Ford Mustang was introduced. (Fitz Fulton Collection)

XB-70 Ship 1 lifts off on its inaugural flight on 21 September 1964. It became the heaviest airplane to fly at a weight of 535,000 pounds at takeoff. The crew consisted of Chief North American Test Pilot Al White, and Prime Air Force Test Pilot Col. Joe Cotton. Ship 1 went on to a distinguished flight test career, and was delivered to the Air Force Museum, now designated the National Museum of the U.S. Air Force, in Dayton, Ohio, on 4 February 1969. (Wings & Airpower Historical Archive)

and steep climb angle, the pilot could not see the earth and had to fly the profile strictly on instruments.

Once the massive acceleration stopped, the pilot was approaching the edge of space and had to switch to RCS, or Reaction Control System, jets to control the rocket plane's attitude. The sidestick controller for the ballistic system was mounted on the left console and superseded the traditional center control stick used for aerodynamic controls. The X-15 pilot could also use another side stick mounted

Detail study of the XB-70's bifurcated air intake. Movable internal inlet ramps controlled the shockwave of air entering the engines; the bomber's folding wingtips were designed to create so-called compression lift, theoretically using shockwaves on the underside of the airplane to increase lift at high speed. This was also called the WaveRider concept. Flight test results on the XB-70's short test program were inconclusive as to whether there really was a benefit. (USAF)

on the right console, which most pilots preferred due to the formidable acceleration forces while the rocket engine was firing.

The X-15's cockpit was designed so the pilot could accurately fly the airplane under unusually high sustained G loads. The need for this accuracy was critical, because at the speeds the X-15 flew, a deviation in flight path angle as small as only 1 degree could mean a difference in peak altitude of more than 7,000 feet. A delay of merely one second in shutting down the rocket motor could mean a difference in speed of 0.5 Mach, as happened when Mach 4 target speeds were first achieved. The actual target had been set slightly above Mach 4, yet the X-15 reached Mach 4.43 on the first attempt.

Pilot tasking to stay precisely "on profile" for the short rocket burn, followed by the extremely high workload during the 3 to 4 minutes of Zero-G flight outside the Earth's atmosphere using ballistic controls, required the skills of the world's best test pilots. Next came preparation for re-entry, which proved to be the ultimate task for any X-15 pilot. His workload remained high throughout the approach and landing, and complete concentration was required literally from touchdown until the airplane came to a complete stop on the lakebed.

Run for the Roses: Turning the X-15 into the X-15A-2

The penultimate flight of the X-15 was an attempt to reach Mach 8 by expanding the rocket burn time to almost 2 full minutes. Temperatures experienced exceeded those for which the airplane was designed, so the decision was made to apply a spray-on ablative coating over the metal skin to protect it from the increased temperatures. Mach 8 was well into the hypersonic regime where an air-breathing engine known as a "scramjet" could provide high velocities at altitudes where there was still atmosphere to support combustion. Since high-Mach speeds would be achieved, a dummy scramjet engine was attached to the X-15's ventral fin to measure scramjet installation temperatures and pressures in flight.

To extend burn time of the XLR99, it was necessary to mount large external fuel tanks on the X-15 with 13,000 pounds of increased fuel and oxidizer weight. Once the external tanks were jettisoned at approximately Mach 2, the original fuel load contained within the fuselage of the X-15 allowed the higher Mach flight. Ablative material developed by the Martin-Marietta company was selected for the coating, which was intended to protect the skin long enough to attain the higher Mach numbers. The coating was duly applied, but the material was bright pink and about the consistency of a pencil eraser. A white outer coating was then applied to "seal" the rubbery pink material and protect it from potential damage.

Test flights were flown with the still-black X-15A-2 carrying the external tanks, to verify successful jettison capability. Subsequently the all-white Ship 2 with ablative materials applied flew carrying the tanks to Mach 4.94 as a full-up systems test below the peak speed intended. The ablative materials seemed to work fairly well, but because the temperatures were lower at only Mach 5 it was not completely certain what would happen at higher Mach numbers when the ablative material would probably start to ablate. One thing that was noticed was the ablative material had charred around the scramjet on the lower ventral fin.

Nevertheless, the ablation layer was considered to be effective, and after retouching burned spots, the next flight was planned to go to peak design speed. On 3 October 1967, Knight launched on what turned out to be not only the fastest flight of the X-15, but the fastest flight of a manned, winged aircraft within the Earth's atmosphere. Initially, the mission proceeded according to what the simulator profile predicted, and after 67 seconds, the airplane was going Mach 2.4 at 72,000 feet and the tanks were jettisoned.

Once clear of both the weight and the drag of the tanks, the aircraft accelerated quickly and Knight leveled off at 120,000 feet and shut down the engine, traveling 6,500 feet per second and flying

This is the YF-12A interceptor photograph issued with President Johnson's announcement of America's new aircraft known as the "A-11." Despite being featured as a double-page spread in Life magazine, this initial photograph and press release were intended to mask the existence of the CIA's super-secret A-12, a most sensitive covert project. The air force interceptor was not intended to exhibit radar stealth characteristics with its additional fins and cylindrical nose, which is precisely why this photo was used. Small conical pods below the wings housed tracking cameras to record weapon launch and separation characteristics. (Tony Landis Collection)

This photo reveals the YF-12's delta wings, twin Pratt & Whitney J58 turbo-ramjet engines, and unique cobra-like forward fuselage. Equipped with a powerful new Hughes radar and armed with the predecessor to the Hughes Phoenix air-to-air missile, the YF-12 would have been a formidable interceptor. Only three YF-12As were built, with the majority of Blackbirds being the SR-71 photo-reconnaissance version. (Tony Landis Collection)

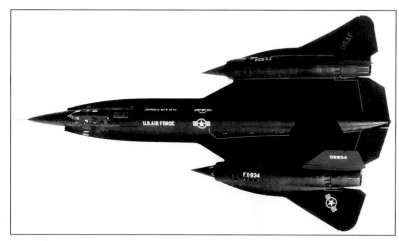

Perfect plan view of the YF-12A shows the unique nose radome and chines astride the forward fuselage. Aft of the cockpits, the SR-71 was identical save for the weapons bays that housed cameras instead of missiles. It is estimated that hundreds of attempts were made to shoot down the Blackbird during its 35-year career, but not one try was ever successful. (Wings & Airpower Historical Archive)

on instruments. But later, analyzed data showed the airplane actually flying 6,600 feet per second at engine shut-down. The engine burned for 140 seconds and radar data showed that the X-15A-2 had attained Mach 6.70 at 102,700 feet, a new (but unofficial) record for a manned rocket aircraft that lasted until 12 April 1981, when Space Shuttle *Columbia* went into orbit.

After the rocket engine shut down, things rapidly began to occur that were out of the ordinary. A "hot peroxide" light came on in the cockpit decelerating through Mach 5.5 that indicated something in the vehicle was overheating. Since there were no chase aircraft available (they were far below the X-15) they could not confirm the problem but later data analysis showed shockwaves at the record Mach number were creating hotspots where they impinged on the lower fuselage. These hot shockwaves were actually burning off the lower ventral fin and the dummy scramjet, and severely damaging the airplane. This problem distracted Knight to the point that when he hit High Key, he was still at 55,000 feet and Mach 2.2, and unsure of the condition of the X-15.

Ground control ordered him to jettison his peroxide, which created a vapor trail to help chase planes find him, but nothing came out and he decided to get the airplane on the ground as soon as

Lockheed YF-12 interceptors were relocated from the CIA's remote secret test location to Edwards and revealed to the press as part of the cover story to conceal existence of the A-12. This is one of three YF-12s making a blistering high-speed pass down Runway 04. (AFFTC/HO via Tony Landis)

The Radar Intercept Officer, clad in his pressure suit, marches out to the YF-12A trailed by the suit technician carrying the portable suit cooler. In May 1965 the airplane set an absolute record over a 15- to 25- kilometer course of 2,070 mph; speed over a 500-km closed-circuit of 1,643 mph; and 1,000-km closed circuit of 1,689 mph. An absolute record for sustained horizontal flight at an altitude of 80,258 feet was also made. Setting these records with an operational aircraft not using its full speed and altitude capability was intended as a purposeful demonstration of American technological superiority over the Soviet Union.

In December 1959, the U.S. Air Force sought to establish a bevy of world records to demonstrate the stunning capabilities of the new "Century Series" family of supersonic fighters. ("Century" stood for the first three-digit designations used on USAF fighter aircraft). The Lockheed F-104 (shown), the Republic F-105, and the Convair F-106 were all chosen for these flights. Later, the ultimate world altitude record for a jet-powered aircraft was set by a rocket-boosted version of the F-104, which reached a peak altitude of 120,800 feet on 6 December 1963. (Wings & Airpower Historical Archive)

This exact aircraft, Republic F-105B Thunderchief S/N 57-5803, set the world's closed-course speed record on 11 December 1959 by flying 1,216.5 mph on the high-altitude course 36,000 feet over Edwards. Piloted by Brig. Gen. Joseph H. Moore, the Thunderchief was flown "clean" without external fuel tanks, and had just enough fuel in its internal tanks to climb to altitude, complete the speed run, and descend for landing on the runway directly below. (Mike Machat Collection)

Brig. Gen. Joe Rogers smiles from the cockpit having just become the fastest jet pilot in the world. Rogers, assigned by the air force to set the absolute speed record in the volley of record attempts in December 1959 piloted Convair F-106A S/N 55-0459 on two blistering straight-line speed runs at altitude over Edwards, and wound up with an average speed of Mach 2.3, or 1,525.9 mph. Maximum speed attained on the first run was Mach 2.41, the fastest a single-engine jet aircraft has ever flown. (Wings & Airpower Historical Archive)

possible. He jettisoned the scramjet but did not feel it depart the aircraft, as he had on the earlier flight. As he went subsonic, conditions seemed to return to normal and he made an uneventful landing on the lakebed. He suspected something was amiss when once the recovery vehicles arrived and the canopy was opened, all ground personnel clustered around the back of the X-15 rather than around the cockpit congratulating him on the flight.

When he exited the cockpit and walked to the back of the airplane he discovered that the scramjet was gone and the lower ventral fin had holes burned through it with signs of melting and skin peel-back. Subsequent review of mission data and radar traces indicated the scramjet had been lost, but fortunately over the Edwards bombing range. Engineer Johnnie Armstrong estimated where he thought it came down, and sure enough, he found the scramjet lying in two pieces on the ground. Examination showed that three of the four bolts used to jettison the scramjet had fired due to excessive temperatures, although it was obvious the scramjet was also badly scarred from excessive heating.

The ablative material had eroded away on numerous areas of the ventral fin, and the metal underneath it had been burned through with major damage to wiring, pressure lines, and the forward compartment housing the peroxide tanks, which were destroyed. Thermocouple temperature readings indicated the ablative coating provided some heat protection for the first 140 seconds of flight (which happened to be approximately when the rocket motor burned out), and by 160 seconds all the ablative material was gone and the metal was unprotected. The ablative material provided adequate protection for the rest of the vehicle, but the ventral fin with its dummy scramjet had obviously created new shockwaves in which temperatures were beyond what was expected.

The Martin-Marietta company, as per contract, removed the remaining ablative material on the X-15A-2 and it was returned to NASA, but never flew again.

It was interesting to note that the X-15 program had been extremely successful in conducting research up to the airplane's limits, but for hypersonic flight, pushing beyond those design limits could be extremely dangerous due to heat. Engineers theorized that had the rocket motor burned for just another 10 seconds, sufficient damage would have been done to the structure to destroy the aircraft. Perhaps it was a warning against attempting to walk into unknown areas without an incremental approach and that at hypersonic speeds, those increments might be smaller than engineers had thought. The one fatal mishap in the X-15 program underscored that lesson, although human factors may have accounted for the accident as much as engineering. USAF test pilot Maj. Mike Adams was killed in that crash when he lost control of X-15 Ship 3 testing an advanced new flight control system.

The X-15's proud legacy was the data it generated in the flight reports on hypersonic flight. The stellar aircraft proved that a space shuttle could successfully return from orbit as a glider and accurately land on a runway. X-15 test pilot Joe Engle, the youngest pilot to ever fly the rocket plane, and youngest to qualify for

Astronaut Wings in the airplane, eventually flew the space shuttle as a NASA astronaut, the only pilot to be dual-qualified in two hypersonic gliders.

There were still records to be set with winged airplanes in the atmosphere. Jet engines were now capable of Mach 2 flight and supersonic cruise rather than brief dash speeds was becoming possible. There were still records to be broken in the Cold War race between the United States and Soviet Union, and perhaps now there was a possibility of commercial jet aircraft carrying passengers at much higher speeds. Although the space program may have been grabbing the headlines, there were still advances in aviation that had more impact on society at large than landing on the moon or orbiting Earth for weeks. The era of supersonic journeys was about to begin.

Supersonic Cruise Becomes Reality: Hustling at Mach 2

With the discovery that the sound barrier was no longer a barrier, the pace of record-setting in the Jet Age picked up speed. Most of these speed and altitude records were being set by fighter-type aircraft while distance records were being set by long-range bombers like the Boeing B-52, but those records were set at subsonic speed. As the Cold War proceeded at a breakneck pace, military forces wanted supersonic bombers as well as fighters, and the first of these new aircraft became reality with the introduction of Convair's ground-breaking four-engine, medium bomber called the B-58 Hustler. The Hustler first flew on 11 November 1956.

Powered by the same afterburning General Electric J79 turbojets that powered Lockheed's F-104 Starfighter and later McDonnell's

With its landing gear tucking-in to the wells, Convair F-106A Delta Dart S/N 56-0459 takes off from Edwards. This is the aircraft initially chosen for the speed record run of 15 December, but last-minute engine problems dictated a replacement F-106A (S/N 56-0467) to be used for the actual flight. (Wings & Airpower Historical Archive)

On 15 December 1959, this Convair F-106 interceptor set a new absolute speed record with its 24,500-pound-static-thrust (in afterburner) J75 engine, compared to the F-104 Starfighter's 15,800-pound-thrust J79. The F-106 with a specially trimmed engine had no trouble attaining an ultimate top speed of Mach 2.41 to become the world's fastest jet airplane. After the YF-12/SR-71 records that were set in the 1960s and 1970s, the F-106 became the world's fastest single-engine jet aircraft, and remains so to this day. (USAF)

F-4 Phantom II, the B-58 was indeed a revolutionary design. Having learned their lesson on the F-102 and F-106, Convair chose a delta-wing design for the B-58. The jet not only used Whitcomb's area rule to improve drag characteristics, but in a novel method to slim down the bomber's fuselage, much of its fuel and the nuclear payload was housed in a large (62 feet long) streamlined shape carried beneath the fuselage, which was released at the target, leaving a very-low-drag aircraft to make its way home with a considerable fuel saving.

It soon became obvious that even with the centerline pod retained, the four J79s, each with 16,000 pounds static thrust in afterburner, could propel the B-58 to Mach 2 where it could fly for almost an hour before having to slow for aerial refueling. SAC commander Gen. Curtis LeMay always preferred long range to high speed, which is why the B-52 was developed to replace the B-47. Nevertheless, more than 2,000 B-47s were built due to the delays in introducing the B-52 into service. The B-58, on the other hand, was intended as an interim supersonic bomber and the introduction of a larger, longer-range Mach-3 bomber called the B-70 eventually replaced it. Only 116 B-8s were built, and they proved to be troublesome during their 10-year operational career because of high landing speeds, reliability problems with the elaborate landing gear installation, and advanced avionics that ultimately proved to be unreliable.

Despite these shortcomings, the B-58 pioneered advanced navigation and bombing systems necessary to accurately hit a target at Mach 2 and an advanced flight control system intended to make extended supersonic cruise as easy as possible for the single pilot by reducing his workload through an advanced autopilot and stability augmentation system. All this effort bore much fruit later with the variable-geometry F-111 and highly maneuverable F-16 fighter that pioneered fly-by-wire flight control systems. Still, the B-58 had an unusually high accident rate for a variety of reasons and nearly one-quarter of the fleet was destroyed in crashes.

The Hustler had a three-man crew (pilot, navigator, and electronic warfare officer [EWO]), with each crewmember having his own individual cockpit. Each cockpit had an ejection seat that was actually installed in a metal clamshell that automatically closed for ejection or in the event of cockpit depressurization; thus, the crew did not have to wear pressure suits, which in the early 1950s were extremely uncomfortable and possibly intolerable for long missions.

The B-58 with its high speed and long range was used to set 19 records (several officially recognized FAI records) during its short 10-year career, including the 2,000-km (1,242-mile) closed-circuit record to 1,062 mph and the 1,000-km (621-mile) record to 1,285 mph on 10 May 1961. Maj. Elmer Murphy received with the FAI trophy

In 1961 the U.S. Navy celebrated the 50th anniversary of American naval aviation by setting records using its new two-seat twin-engine McDonnell F4H-1 Phantom II. These included absolute speed at 1,607 mph, absolute height of 98,556 feet, and a new record for speed at low altitude of 903 mph using the old 3-km speed course. Flying supersonic at such low altitudes was extremely dangerous. During an earlier practice flight for the 3-km course record, stability augmentation system failure caused a severe pilot-induced oscillation, resulting in over-stressing the aircraft causing airframe disintegration with loss of both pilots.

established by Louis Bleriot in 1930 for presentation to the first pilot to exceed 2,000 kmph (124 mph) for a continuous period of 30 minutes. Two weeks later, Maj. William Payne flew from Texas to Paris setting enroute record times of 3 hours and 40 minutes from Washington and 3 hours and 20 minutes from New York to Paris. (Unfortunately that Hustler crashed during the Paris Air Show with the loss of the entire crew.)

One extremely long-distance record was set flying 8 hours and 35 minutes from Tokyo to London at an average speed of 938 mph, which included slowing to Mach 0.8 for five aerial refuelings. The actual cruise speeds for more than half the trip ranged from 1,350 to 1,400 mph, making the flight the longest supersonic nonstop flight in history.

Despite its own record of having set the highest number of records ever held by one type of aircraft, the B-58 was on its way out. Beginning in December 1969, the entire fleet was retired to the "boneyard" at Davis-Monthan AFB in Tucson, Arizona. Changes in the Soviet threat environment with increased reliance on missiles meant that the era of higher and faster jet airplanes was coming to an end. The B-58's replacement was the General Dynamics (formerly Convair) FB-111 with variable-sweep wing. Later models of the FB-111 were capable of more than Mach 2.2, but they never went that fast because their mission was to fly at high subsonic or low supersonic speeds at low altitude "under the radar" to avoid

Soviet surface-to-air missiles. The quest for higher speeds, which could only be achieved at high altitudes, seemed to be over.

In 1964, however, an announcement was made by U.S. President Lyndon B. Johnson about a new airplane capable of cruising at Mach 3 at altitudes far above the B-58's capabilities. This airplane was advertised as being a U.S. Air Force aircraft, but in reality its original customer was not a military service at all, but rather a government agency that became known to the world by its cryptic three-letter title: CIA. New speed and altitude records were going to be set, but these new airplanes were neither fighters nor bombers, and in fact, carried no weapons whatsoever. They were exceedingly important, however, and helped keep the Cold War from becoming hot.

The Black World of High-Altitude Reconnaissance

The atomic bomb and increasing tension between the United States and Soviet Union made development of photographic intelligence even more critical in the post–World War II era. The newly formed Central Intelligence Agency (CIA) was tasked to find out if the Soviets had a long-range bomber fleet that could deliver atomic bombs. The CIA determined that a manned reconnaissance airplane was necessary to conduct covert operations over the Soviet Union

The McDonnell Douglas F-15 Eagle first flew in 1972, and was designed from scratch to be the world's most outstanding air superiority fighter, capable of dominating any airspace in air-to-air combat. Learning from the hard lessons of Southeast Asia, where U.S. fighter aircraft racked up a kill ratio of only 3-to-1, the F-15 featured smokeless engines, a large 360-degree-vision bubble canopy, and engines that were separated below the wing. As a result, the F-15, which is flown by many air forces around the world, has amassed a kill ratio today of 144-to-0. (Mike Machat)

during the Cold War, and the CIA contracted with Lockheed to develop a high-altitude single-seat reconnaissance aircraft. Given the bogus designation of U-2, this airplane was capable of flying over the Soviet Union at 70,000 feet, higher than any anti-aircraft gun or fighter of the early 1950s could reach. More amazing was that with its single large camera, it could photograph small airplane-size targets from that high-altitude.

The air force assisted the CIA in the development and training of a special hand-picked cadre of U-2 pilots, and ordered dozens of U-2s for their own use as a high-altitude reconnaissance platform. The prototype first flew in 1954 and proved to be very good at high-altitude, but could only cruise at approximately Mach 0.7. Therefore, the U-2 relied on its height advantage and electronic countermeasures for survivability. It was no accident that the Soviet Union set several highly publicized and officially recognized altitude records using modified air defense jet fighters from 1959 through 1962, reaching altitudes from 69,000 feet to 113,000 feet (the latter was set by an "E-66," the MiG-21 Fishbed with rocket assist). These records were a warning to the American intelligence community that the U-2's days were numbered.

The downing of a CIA U-2 in 1960 by surface-to-air missiles over central Russia with the hardly credible cover story that the pilot made a navigational error, was soon exposed when the pilot was produced

by the Soviet government and put on a very public trial for being a spy. The wreckage of his airplane, complete with camera and film, became Exhibit A in his trial and later in a museum where it still sits today. Pilot Francis Gary Powers was found guilty and was convicted, but was later exchanged for a convicted Soviet spy and returned to the United States. Less widely known was that the United States and Soviet Union signed a secret agreement that the United States would never again fly a manned spy plane over Soviet territory.

But the U-2 had actually done its job, having flown 25 missions over Russia and confirming that the Russians indeed did not have a large fleet of jet bombers with which to launch a large-scale atomic attack on the United States. The U-2 remains in service on its 60th birthday in the 21st Century, still flying with the air force and NASA. In 1989, when early U-2Cs were taken out of service with NASA, the aircraft set several official horizontal altitude records (73,260 feet) and time-to-climb records for a single-engine jet aircraft before it landed to enter a museum. It is the highest flying single-engine jet aircraft in aviation history.

While Powers was on trial in the Soviet Union, the U-2's successor was busily being developed by engineers at the Lockheed Skunk Works. It never flew over the Soviet Union thanks to the secret treaty, but it turned into what many people regard as the ultimate

It was not until 1967 that the Soviet Union publicly revealed its entry for an airplane to counter the Mach-3 YF-12, and records were set using the new MiG-25 Foxbat (also known by its "civilian" designation, E-266). Speed records over 500- and 1000-km closed-circuit courses were 1,853 and 1,815 mph, respectively. In 1977, new absolute records for height were set of 123,523 feet and 115,584 feet, with payloads of 1,000 or 2,000 kg, respectively. With the MiG-25 being an interceptor, those altitudes were reached in zoom climbs. The Foxbat was capable of reaching Mach 2.85 but had insufficient fuel to remain at that speed for very long. (USAF)

masterpiece of advanced aeronautical engineering that even today would be hard to duplicate. It had many names, such as Oxcart, Cygnus, Archangel, U-3, A-12, SR-71, and even RS-71 (briefly), but to the majority of the world it will always be known by the sinister name "Blackbird."

By 1959, the CIA knew the U-2 was becoming vulnerable to Soviet air defenses and began to analyze the problem. They knew that fighters as well as surface-to-air missiles would be able to reach the U-2 even while cruising at 70,000 feet and shoot down the unarmed and subsonic reconnaissance airplane. The CIA began working to develop a U-2 replacement that could fly 20,000 feet higher at 90,000 feet and cruise at 2,000 mph (Mach 3) rather than at subsonic speed. As long as they were specifying a new aircraft, they also decreed that

it should not be readily visible to radars, because the Russians had tracked the U-2 on radar on all its supposedly covert missions.

By 1958, Lockheed was conducting studies on designing a replacement to satisfy those requirements. A number of designs were developed under the designator letter "A" (unofficially standing for Archangel because the U-2 had been known as the Angel) using turbojets, ramjets, and in one case, even an inflatable airplane so it wouldn't show up on radar. More practical designs resulted, and by 1959, a proposed twin-engine turbojet design known as the A-11 appeared that could meet the speed, height, and range requirements. In summer 1959, the A-11 was reconfigured as the radar stealthy A-12, a redesign that made it more resemble a spaceship than a jet fighter. The air force ordered 30 two-seat versions of the A-12, later designated "SR-71," while keeping from the public the existence of the super-secret single-seat CIA A-12, which the SR-71 later replaced.

In 1975 a specially modified McDonnell Douglas F-15 Eagle broke the MiG-25's time-to-climb records. The F-15 was stripped of 1,800 pounds, including its paint as well as the gun, flaps, speed brake, and actuators. Only enough fuel to reach the record altitude was loaded aboard. Replacing the jet's tailhook was a tie-down bar equipped with an explosive bolt so the pilot could plug-in both afterburners while stationary on the runway, then press a button releasing the airplane and starting the clock. On 1 February, weather conditions were perfect and a time-to-climb record of 208 seconds to 30,000 m (98,520 feet) was set by the F-15 "Streak Eagle," beating the MiG-25's records by 36 seconds. (USAF)

Cruising at Mach 3: The XB-70 and Blackbird

At the same time the CIA was developing its secret A-12, the air force was working on a Mach-3 long-range bomber. The design effort on the North American XB-70 Valkyrie began in the mid-1950s, but the first prototype did not fly until 1964. When it made its first flight, it was the heaviest airplane in aviation history at 490,000 pounds, yet it was also supposed to be the fastest, powered by six General Electric J93 afterburning turbojets at 30,000 pounds of thrust each. Since supersonic lift-to-drag (L/D) ratios were lower than subsonic designs by approximately 50 percent, according to the Breguet range equation there would be difficulties in the design achieving long-range at high speeds.

In the late 1950s, NACA developed a theory called "compression lift" that stated with proper shaping of the wing and fuselage of a high-speed vehicle, the shockwave on the underside of that vehicle would increase the lift considerably so that a higher L/D might be achieved by "surfing" on that shockwave (also known as the "Wave-Rider" theory). This theory was actually derived for ultra-high-speed flight with the X-15 or ICBM reentry vehicles, but it was thought that at Mach 3 it might be possible to achieve a similar effect by folding down the wingtips of a bomber.

The XB-70 Valkyrie was built with folding wingtips, and the wing was so large that each folding tip equaled the entire wing area of the Convair B-58. The airplane was made of titanium and a stainless steel honeycomb, both of which could withstand heat. The honeycomb was intended to reduce the weight of stainless steel to allow long range, but proved to be extremely difficult to fabricate and weld, which delayed the program for more than a year.

As this was taking place, however, the development of intercontinental ballistic missiles suddenly rendered manned bombers obsolete in the view of some American officials, and the B-70 program, as impressive as the airplane looked, was canceled before the first prototype ever flew. Two experimental prototypes were contracted, however, to study the properties of large supersonic aircraft in anticipation of America's SST program. Much of the test program was devoted to gathering data on sonic booms and air traffic control procedures at high altitudes and high Mach for these giant transports that were proposed for the 1970s.

Both Valkyries were flown at Edwards Air Force base until June 1966 when the second and more advanced aircraft was lost in a mid-air collision. The remaining bomber continued flight test with NASA and completed the minimal 128-hour test program. That airplane is permanently displayed at the National Museum of the Air Force in Dayton, Ohio. The XB-70's total flight time at Mach 3 during the entire test program was less than two hours, so very limited data was collected for the supersonic transport.

A flight test results paper published at the end of the program made several comments on the difficulty of controlling such a large airplane at Mach 3 at 70,000-feet altitudes, but it also noted that range estimates on the test aircraft seemed to be deficient by almost 20 percent, which made one wonder whether the compression lift theory would actually work for a vehicle traveling at "only" Mach 3.

Lockheed's Magnificent Blackbird Takes Center Stage

While the impressively large all-white (for nuclear flash protection) XB-70 was being constructed in Palmdale, another, smaller

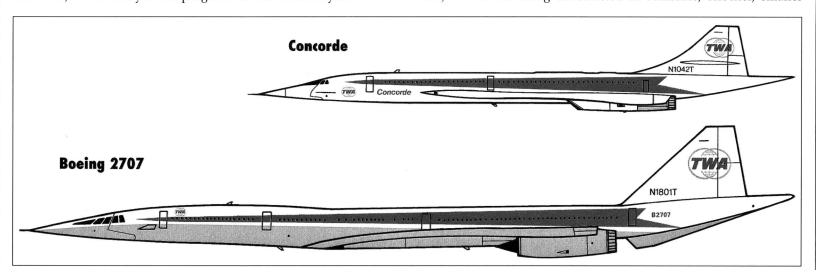

Boeing won the SST design competition with its mammoth 2707 that would cruise at Mach 2.7 and make two round trips across the Atlantic in a single day. The proposed passenger load for optimum economics was nearly 300 passengers carried in an extremely narrow fuselage over 200 feet long. The noisy high-thrust turbojet engines necessary to propel this aeronautical monster created potential environmental and noise abatement problems that led to the jet's cancellation. Seen here in hypothetical TWA markings, the 2707 is compared to the Concorde even though that aircraft never flew for that airline. (Mike Machat)

In this dramatic painting by R. G. Smith, a McDonnell Douglas SST (Supersonic Transport) proposal from the 1980s cruises well above Mach 2 at 60,000 feet. Two decades earlier, Lockheed and Boeing were finalists competing to design an American SST with configurations intended to cruise either slightly above Mach 2 or at Mach 2.7. Different engineers at the different companies favored different configurations such as a radical 60-degree swept wing, which was highly advantageous at a Mach 2+ cruise, but had serious challenges in slowing down enough to land. (Mike Machat Collection)

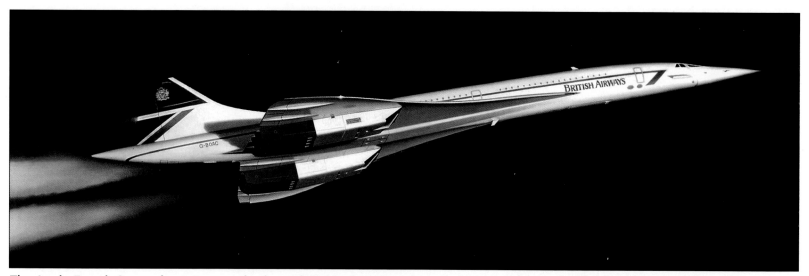

The Anglo-French Concorde project was both a political and engineering wonder. Beginning as a design study in the early 1960s, Concorde first flew in March 1969 and inaugurated the world's first passenger service in 1976. Flying at a cruise speed of Mach 2.04 at 60,000 feet, Concorde became the fastest airliner in the world setting numerous speed records in the process. Passengers flying westbound landed in New York two hours before they took off from London or Paris (according to local time). (Mike Machat)

vehicle was being built initially at Lockheed's Burbank facility, with final assembly later moved to Palmdale. The exotic SR-71 Blackbird was initially the "silver bird" because the titanium of which it was constructed is an extremely shiny metal. The initial flight of the first A-12 was conducted at a remote desert test location in April 1962, and the flight almost ended in a disastrous crash landing due to an improperly located center of gravity.

A second flight attempt was more successful until pieces started falling off the airplane, but fortunately, they were all noncritical structural panels. The pieces were duly repaired and the third time was the charm when, on 30 April, the first Blackbird flew successfully for 59 minutes, concluding its first test sortie with a low-altitude gear-down fly-by over the official party of engineers, technicians, and observers.

The drive toward cruising at more than Mach 3 began in March 1963 with delivery of the aircraft's definitive Pratt & Whitney J58 engines. Numerous inlet/engine problems began to appear above Mach 2, which had to be corrected, and the inlets of the Blackbird wound up being modified in ways the initial design had not accounted for. One of the biggest problems was known as an inlet "unstart" caused by mispositioning the normal shockwave inside the engine inlet, and it wasn't until November 1963 that the A-12 was able to reach its design Mach number of 3.2.

On 3 February 1964, the first sustained flight with speeds above Mach 3.2 for 15 minutes at an altitude of 84,000 feet took place piloted by Lockheed test pilot Jim Eastham. The A-12 test program continued in secret as the CIA measured performance of its new airplane and prepared to deploy it operationally. The A-12's range was approximately 3,000 miles between aerial refuelings, and it was soon obvious that to flight test the airplane it was necessary to move it from the black, covert world to a more public status.

On 29 February 1964 President Johnson announced what he called the A-11 as a Mach-3 jet interceptor, which was to be based at Edwards Air Force Base. The A-11 was actually the Air Force's YF-12 interceptor, and two of them were duly flown to Edwards from the remote flight test base in Nevada and eventually were demonstrated to the press. The air force announced it was building a two-seat reconnaissance version of the YF-12 to be known as the SR-71. The resemblance of the A-12 to the SR-71 was so strong that the CIA decided to paint their aircraft like the SR-71 to deceive unauthorized observers.

The CIA turned over its high-Mach air-breathing reconnaissance mission to the air force in 1968 as well as the U-2 mission in 1974. The SR-71 remained in service until 1990 with the final SR-71 flight being made by a NASA crew at the Edwards AFB air show on 9 October 1999. During the SR-71's operational career, the fleet (which never exceeded 12 aircraft on operational status at one time) amassed a total of 11,675 hours of flight time at more than Mach 3. A total of 29 SR-71s were produced with 12 being lost in accidents. Despite more than 1,000 alleged intercept attempts, none were ever brought down by enemy action. Kelly Johnson referred to the Blackbird program as the most challenging engineering project he had ever experienced because nothing could be purchased off the shelf and everything had to be purpose designed due to the thermal environment of sustained Mach 3 flight.

In 1965, the air force used the YF-12 to establish the absolute speed record over a 15- to 25-km course of 2,070 mph, and a sustained altitude in horizontal flight record of 80,258 feet. Additional records were set for speed over 500- and 1,000-km closed-circuit

Russian Tupolev Tu-144 SST first flew in December 1968, but never enjoyed the Concorde's operational success. Seen here is a later version on loan for a NASA research project. The author flew air force interceptors in West Germany in 1973, and caught sight of an airliner flying directly below him, which he initially identified as a Concorde until he saw the Russian flag on the tail. His flight turned around and shadowed the airplane until they realized it was en route to the Paris Air Show. The two crews bemoaned the fact that nobody had a camera to take a picture of it, as it was a gorgeous airplane in flight. Tragically, one week later, that was the Tu-144 that crashed at the Paris Air Show. (NASA)

courses of 1,643 and 1,689 mph, respectively. These records deliberately did not represent the SR-71's full operating envelope, but in 1976, SR-71s flown by operational crews in operational configurations set the new speed record of 2,193 mph, sustained height in horizontal flight of 85,069 feet, and speed over a 1,000-km closed-circuit course of 2,092 mph. These records remain unbroken to this day, and what some consider the premature retirement of the airplane in 1990 closed the book forever on the world's fastest and highest flying jet aircraft after more than 25 years of service.

USA versus USSR: Speed and Altitude Records in the Cold War

Between 1958 and 1962, speed and altitude records exchanged hands many times in a competition between the United States and Soviet Union tailed, as well as between the air force and navy. The Soviet Union used delta-wing fighters to demonstrate speed, height in horizontal flight, and absolute height, which could be achieved in a zoom climb that never leveled off at peak altitude. These aircraft were the mainstay of Soviet air defense forces and records were intended to publicize their capabilities against American bombers.

Soviet records included 500-km closed-circuit speed of 1,452 mph (Sukhoi Su-11), absolute height of 94,659 feet (Su-9), and height in horizontal flight of 69,456 feet (Su-11). The E-66 (a non-military designation for the MiG-21) reached an absolute speed of 1,484 mph and absolute height of 113,891 feet. The E-66 record set by Col. Mossolov in 1959 was the first absolute speed record ever held by the Soviet Union, but it was certainly not to be the last.

In 1962, the absolute record for speed was raised to 1,666 mph by Col. Mossolov flying an airplane known to the FAI as the E166, which basically looks like a MiG-21 on steroids using a single engine with 22,000 pounds static thrust in afterburner (compared with 12,000 pounds thrust of a normal MiG-21 engine in afterburner). The E-166 was intended as an experimental prototype and never entered service, its fuselage being 66 feet long versus the MiG-21's 44 feet. When the airplane was put on display at the 1967 Moscow Domodedovo air show the overwhelming impression was that of a monster engine that had an airplane attached to it, somewhat like the Gee-Bee racers of the 1930s.

The USAF used a wider variety of airplanes to set records during that same period using the single-engine F-104A with its tiny wing to set the pace with a 1,404-mph speed record and absolute altitude record of 91,243 feet set in 1958. In December 1959, a new absolute speed record was set by a delta-wing Convair F-106 interceptor with its more powerful Pratt & Whitney J75 turbojet producing 24,500 pounds static thrust in afterburner, raising the record to 1,526 mph. In 1961 the navy celebrated the 50th anniversary of American naval aviation by setting a series of records using its new twin-GE J79-powered Mach 2+ McDonnell F4H-1 Phantom II for an absolute speed record of 1,607 mph. A record for absolute altitude of 98,556 feet was also set by a production model F-4B Phantom II.

In 1959 and August 1960, a new record for speed at low altitude of 903 mph was set using the old 3-km speed course rules of remaining 100 meters (330 feet) above the ground. Such supersonic speeds at such a low altitude were extremely dangerous and, in fact, in an earlier practice run for the record, a stability augmentation system failure in an F-4 resulted in a pilot-induced oscillation on the 3-km course, causing an overstress and destruction of the aircraft with the loss of both pilots. This was to be the last attempt to set at low-altitude jet speed record until 1977 when civilian test pilot Darrell Greenamyer flew an F-104 Starfighter named "Red Baron." The jet had been scrounged together from cannibalized and "acquired" parts and engines to set a record of 988 mph over a 3-km course in Nevada.

It wasn't until 1967 that the Soviet Union revealed its entry for an airplane to counter the air force's Mach-3 YF-12, and records were set using the new twin-jet, twin-tail MiG-25 Foxbat (also known as the E-266). Records set for speed over 500- and 1,000-km closed-circuits were 1,815 and 1,853 mph, respectively. New absolute records for altitude were set in 1977 of 123,523 feet, and with payloads of 1,000 and 2,000 kg, to 115,584 feet.

The YF-12 and SR-71 were both capable of extended periods of cruise at Mach 3 and above, whereas the MiG-25 could perform a zoom climb trading airspeed for altitude briefly, which was the method used to achieve absolute altitude records. Sustained flight

Saving the Crew at High Speed and High Altitude

As military aircraft spent more time at high speeds and also high altitudes, concerns grew as to how to save the crew in the event of an emergency. The first use of an ejection seat at supersonic speed occurred in an F-100 Super Sabre where the pilot ejected at approximately Mach 1.05 in a high-speed dive and was killed by the wind blast, which was the equivalent of 500 mph at sea level. Supersonic jets such as the B-58 and XB-70 would be spending long times at high-altitude and high-speed and it was proposed that their escape systems be converted into capsules to both protect them from wind blast and also allow the pilots to operate in a shirtsleeves environment rather than a pressure suit. When the capsule became enclosed prior to ejection, it provided a pressurized environment for the pilot. During ejection, the pilot remained protected inside his capsule, which had its own parachute and air bags on the bottom to cushion the landing.

The B-58 was the first operational aircraft to have a capsule, and it was tested at Edwards AFB using tranquilized bears strapped into the capsule seat in lieu of humans. This seat capsule concept actually seemed to work, although the seat was complicated and heavy, and in the event of an ejection most touchdown were best described as controlled crashes. The XB-70 used a similar concept, but the one time its escape capsule was used, one of the pilots was unable to eject due to the complexity of the capsule system and the other pilot left his boot prints in the aluminum floor of the capsule because of the severity of the landing's impact. He never flew again.

The operational two-man F-111 low-altitude supersonic penetrator aircraft could fly 800 mph (more than Mach 1.5) at sea level, and ejection seats seemed completely impossible for crew survival at those speeds. Supersonic stability of an ejection seat at low altitude was also in question, so the concept was to take the entire cockpit with the two-person crew during ejection and have massive parachutes and airbags deploy to lower the entire cockpit escape capsule to the ground. An extensive development effort with very powerful rocket motors and explosive wire was used to validate this concept, which was used successfully many times throughout the F-111's career.

The Rockwell B-1A bomber was originally intended to also use a similar cockpit capsule, only in the case of the B-1 with its

Douglas Master Artist R. G. Smith depicted a high-speed, high-altitude escape system that used the entire forward fuselage to protect the pilot during bailout. Here, as the escape pod is being jettisoned from the disabled aircraft, the pod is initially stabilized by the small drogue chute. The canister in the middle of the drogue chute rises then deploys a large full-canopy parachute at lower altitudes to return the capsule to Earth with the pilot seated safely inside. (Douglas Aircraft Company)

four-man crew, the capsule had to be considerably larger than the F-111's. During rocket sled tests using mockups of the B-1A capsule, it was discovered that above 300 mph, it was impossible to keep the capsule in a stable attitude. As a result, the large parachutes did not open properly and ejection ended in failure. After a fatal accident during early B-1 flight testing, the decision was made to give each crewmember an individual ACES II ejection seat, and the remaining 100 production B-1B aircraft were so equipped.

For the Blackbird series, Lockheed elected to rely on the pressure suit and a Lockheed-built ejection seat for crew egress,

Illustrating the slightly different concept of using a clamshell pod to enclose the pilot before firing him out of the aircraft, the escape pod of the Mach-2 Convair B-58 Hustler is being ejected from the bomber during high-speed taxi tests at Edwards AFB. Inside the pod a tranquilized brown bear was used to simulate human physiology for these high-risk tests. Later ejections were performed successfully at altitude with a human subject. (USAF)

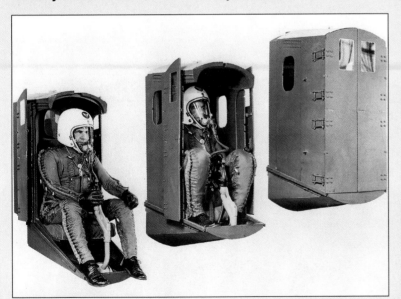

Actually built, but never operationally tested, this ingenious but highly-complex escape device was referred to by doubting engineers as "The Phone Booth." Intended to protect pilots during high-altitude and/or high-speed bailouts, this apparatus had spring-loaded doors that would snap shut in a split-second before the "booth" was ejected from a crippled aircraft. It is assumed the pilot would open the doors and manually bail out from the capsule at lower altitudes, but no provisions for parachutes of any sort are indicated in these photographs. (Republic Aviation Corporation)

Rockwell International's B-1A prototypes were fitted with a single large escape pod for all four crewmembers who would be ejected from the aircraft and returned to the ground by large parachutes. Inflated gas bags at each lower corner would then cushion the landing shock to ensure survivability. In the single case where this system was used to escape a crashing B-1A, one pilot was killed and the other seriously injured when one of the shock-absorbing bags did not inflate. Production B-1Bs were then fitted with McDonnell Douglas ACES II ejection seats for each crewmember. (Wings & Airpower Historical Archive)

When all is said and done, a simple, efficient, and well-designed ejection seat is the safest and most reliable means of providing escape from a stricken aircraft. Although this seat was designed in the late 1950s, it offered near Zero-Zero capability with an "initiation to fully inflated parachute" interval of less than 3 seconds in Mode-1 (low altitude). Operational seats today have reduced that time to 1.8 seconds, and can save a pilot from an inverted aircraft flying as low as 300 feet. (Republic Aviation Corporation)

reasoning, as did the X-15 program, that most ejections at high-altitude had minimal wind blast, known as "Q," due to the thinness of the atmosphere at high altitude. A properly designed ejection seat with adequate restraint would provide the wind blast protection, and the suit's life-support system would be adequate for the 5 to 6 minutes of free-fall from 80,000 feet down to 13,000 feet where normal man-seat separation and sub-sequent parachute deployment at 10,000 feet would occur. The ejection seat had a further advantage in that it could be used in "zero-zero" conditions at low altitude during the hazardous heavyweight high-speed takeoff. Over a dozen SR-71 aircraft out of the 30 built were lost, but all crewmembers that used ejection seats were recovered, giving the Lockheed seat a 100-percent safety record.

at Mach 3 was impossible for the Foxbat, as its normal top speed was somewhat below Mach 3. Much Western-world curiosity was evident when the MiG-25 was first shown at the 1967 Moscow air show, its huge engines and massive size attracting considerable attention.

Back in the United States, the McDonnell Douglas F-15 Eagle air superiority fighter the air force was developing to replace the Phantom exhibited many MiG-25 design features, but it was uncertain what the top speed of the Foxbat really was. The air force now chose to set new records not so much in traditional competition with the navy, but to entice the Soviet Union to reveal more of the MiG-25's true capabilities.

In 1976, as part of America's bicentennial celebrations, an air force SR-71 took most of the level-flight altitude and sustained speed records away from the MiG-25, with the exception of the absolute height records, at which the Soviet interceptor excelled in zoom climbs. In February 1975, however, an F-15A was modified with weight reductions totaling almost 2,000 pounds, including stripping off all the paint on the jet and removing all excess equipment. By never flying with a full load of fuel, this airplane was successfully used to break the MiG-25's time-to-climb records.

The penultimate record time to 30,000 m (98,430 feet), and for that and the 20,000-m (65,620-foot) records, the pilots needed to wear pressure suits, which are not normally installed in an F-15. More than 60 flights were flown practicing the profiles to minimize the time required to reach the altitudes, and the flights were made in winter from Grand Forks AFB, North Dakota, because cold temperatures produced more thrust in the F-15's Pratt & Whitney F100 turbofan engines. It was calculated that to achieve the best time to 30,000 m, the temperature at altitude should be 7 degrees C, which is colder than standard, and the F-15 team had to wait several days for that temperature to occur.

Part of the drills involved practicing dead stick landings with the F-15, because it was assumed the engines would flameout at the high altitudes. In the event they could not get an engine relighted, they might have had to execute a glider-type landing, which in North Dakota meant they had to land at the civilian airport in Fargo. On 1 February, weather conditions were correct and a blazing time-to-climb record of only 208 seconds to 30,000 m was set by the F-15, beating the MiG-25 record by 36 seconds.

The F-15 actually topped out at 103,000 feet in a ballistic profile similar to the X-15, but no absolute altitude record was claimed because that record had already been set in 1973 at 118,898 feet. In 1977, the MiG-25 raised that mark to 123,523 feet. Then the F-15 Streak Eagle's records were all beaten by a Soviet prototype of the Sukhoi Su-27 Flanker, but the Cold War race for records ended a decade later when the Soviet Union was dissolved and the threat of global nuclear war (which triggered the need in homeland defense for higher and faster aircraft in the first place) completely disappeared.

The SST: Supersonic Dreams versus Subsonic Reality

The rapid success of the jet airliners starting in 1958 led to the logical desire for airliners that could cruise at supersonic speeds. By 1962, Britain and France had decided to combine their efforts to produce a delta-wing supersonic airliner carrying 100 passengers at Mach 2. Four turbojets of approximately 35,000 pounds thrust each allowed this airliner to cross the Atlantic in 3-1/2 hours, or half the time of subsonic jetliners. It was estimated there would be a world market for 300 or more of these airliners. America's aviation industry recognized this as a challenge to Boeing and Douglas' market domination, so interest began to grow in the United States to build an airplane both larger and faster than the European design. Within a year after committing the United States to go to the moon before the decade was out, President John F. Kennedy also promised the federal government would back the development of a supersonic transport.

Studies by the FAA, put in charge of the project assisted by NASA, looked at two different designs: one to cruise at Mach 2 similar to the European SST, the other designed to cruise at Mach 2.7 carrying more than 200 passengers. Studies done by Boeing and Lockheed on the economics of both designs led to the conclusion that the Mach 2.7 giant was preferable for supersonic travel. NASA engineers developed a number of SST proposals, and because they were on the source selection board, the competing companies had some idea of what was desired. Lockheed produced a pure delta-wing design using a "cranked delta" (80 and 60 degrees) sweep and relied on their valuable experience building the Blackbird with its cruise speed of Mach 3.2 rather than Mach 2.7.

Boeing competed in the TFX variable-geometry fighter competition and used that experience to produce a design with variable-sweep

Boeing's 747 revolutionized global air travel, and although the popular press called it a "jumbo jet," a more correct description was wide body. Wide-body airplanes have two aisles, which allows up to 10 seats per row. The wide fuselage is key to carrying more cargo and fuel in addition to more people. Japan Airlines operated high-density 747s carrying more than 500 passengers. When introduced in 1969, the 747 carried more passengers than had ever been flown in a single airplane, as well as more fuel than ever before, thus extending the range of nonstop flights over distances never before achieved. (Mike Machat Collection)

wings, although it was estimated by Lockheed that would add 50,000 pounds to the weight of the SST. The wings of the Boeing design, when fully swept, resembled some of the NASA SST designs, but then extended nearly straight to land. Both airplanes were huge (200- to 250-feet-long fuselages), weighing more than 500,000 pounds each and requiring four 60,000-pound-thrust engines.

On 31 December 1966, the FAA announced that Boeing had won the competition, along with General Electric for the engine. Now the serious design work began, but in the SST race, the newly named European Concorde was now only three years away from its first flight. Deep in the Soviet Union, a Russian supersonic transport was also being developed and was only two years away. The European and Soviet delta-wing designs looked almost identical, and although the Tupolev Tu-144's official NATO code name was Charger, it was unofficially known to everyone as the "Concordski."

The Boeing SST design turned out to be far too heavy and Boeing was forced to accept a fixed-wing delta concept with a conventional tail for its prototype, which was never built. The U.S. Senate then defunded the SST as part of the country's post–Vietnam War emphasis on the environment, rather than aerospace. It took more than six

years for the U.S. Supreme Court to allow the Concorde to operate into New York due to local jet noise objections. Europeans regarded this as a classic example of Americans saying, "If we can't have one, you can't either."

The maiden commercial flights of the Concorde departed simultaneously from London and Paris in September 1976, but their destinations were Bahrain and Rio de Janeiro, due to U.S. political strife. Meanwhile, the Senate, in deference to environmental concerns, had forbidden supersonic travel by commercial aircraft over the United States. This, combined with the first "fuel crisis" resulting from the Yom Kippur War in the early 1970s, resulted in much higher fuel prices and spelled doom of the supersonic transport, even at Mach 2.

Only 16 operational Concordes were built for the British and French national airlines, as a matter of prestige, but their high cost and limited destinations relegated them as commercial failures. The Soviet Concordski fared no better, only flying freight and mail between Moscow and Siberia with limited passenger seating. Several Tu-144 crashes, including the one at the Paris Air Show in 1973, brought back memories of the original Comet jetliner experience, and in the year 2000, a disastrous and well-filmed fatal Concorde crash taking off from Paris spelled the end of supersonic passenger flight when the Concorde was grounded for two years. In 2003, during the centennial of aviation, the last French Concorde landed at the Paris Air Show to be enshrined in a museum, and supersonic transatlantic passenger flights (at $6,000 one-way prices) were now a thing of the past.

While Boeing was struggling to build its SST, another Boeing gamble was occurring that would revolutionize jet travel. That airplane was the Boeing 747 and it was the first of the jumbo jets, or wide bodies. Based on Boeing's experience to acquire a contract for the air force's giant new C-X cargo transport (known later as the C-5), its commercial aircraft division decided it would be possible to build a much larger jet airliner than ever before attempted. This goliath of the sky would have two aisles and a cabin width containing up to 10 seats across. It would carry as many as 500 passengers in a high-density configuration on the main deck, although 350 to 420 passengers would be standard.

A second, shorter upper deck behind the cockpit was added, initially providing additional VIP seating and a lounge, although the actual aerodynamic reason was to augment the fuselage cross-section in accordance with Whitcomb's area rule, allowing economical cruise speeds up to Mach 0.9. It would be propelled by four large turbofan engines producing 40,000 to 50,000 pounds of thrust with a predicted profitability breakeven point of 400 airliners. Steady progress on the 747 encouraged the airlines to purchase the airliner, and when the first airplane was rolled out in September 1968, firm orders for 158 were already on the books.

Ten years later, 724 original "747 classic" airplanes had been delivered. The airplane with its distinctive humpback appearance was known worldwide, including a new presidential version known as Air Force One. The next-generation 747-400 incorporated a two-pilot "glass cockpit" with numerous improvements, including refined aerodynamics and even more powerful jet engines allowing increased weights. This also allowed longer-duration flights linking cities nonstop that had never been connected before. A 747-400 flight from London to Sydney, Australia, set a new distance record for commercial airliners of 11,156 miles in 20 hours and 8 minutes, a record that still stands today. That distance, incidentally, is approximately halfway around the world.

The 747 success in the development of high-bypass-ratio turbofans led to the development of further wide-body aircraft using only three engines and then later two. Although not having the volume of the 747, a tri-jet wide body with fewer seats could be profitable on shorter routes. The McDonnell Douglas DC-10 and Lockheed L-1011 soon joined the 747 fleets, and eventually Boeing adopted that philosophy with the twin-jet 767 in 1982. The same engines used to power the larger 747 could also be used to good advantage on smaller wide bodies, which increasingly became the trend as mass travel expanded all around the globe. The fact that the high-bypass turbofans were also much quieter than turbojets was also a plus.

Although a Concorde flown by a French crew holds the FAI commercial record for around-the-world flight set in 1995 with an average speed of 811 mph (and six refueling stops along the way), in the competition between supersonic and transonic jet airliners, the long-range high-capacity 747 proved to be the best solution. It is estimated the 747 has carried more passengers than the entire population of the Earth since it entered service. It led to the development of more powerful high-bypass turbofan engines as well as cheaper tickets so that more people could fly to distant destinations than ever before. The SST and Concorde experience indicated that higher speed was not a necessity for future jet travel.

Now that commercial aviation had advanced into being essentially the aerial component of mass transit, another aspect of aerospace technology was about to revolutionize not only aviation, but also the everyday life of virtually everybody on this planet. It was the digital revolution that gave us the computers and digital devices that we use and take for granted today. Most people do not realize that, essentially, it all started with fulfilling the needs of the aerospace industry.

This popular publicity photo of a flight attendant standing in the inlet of a high-bypass turbofan illustrates the other key to the success of the 747, engines with 50,000 pounds of thrust that did not require noisy and fuel-gulping afterburners. The Pratt & Whitney JT9D and General Electric CF6 engines originally used for the 747 led the way for even higher-bypass turbofans that today produce thrust of as much as 100,000 pounds. This, in turn, allows today's wide-bodies to fly with only two engines instead of four, and with ultra-long-range capability that can link any two cities in the world.

Chapter 7

THE DIGITAL AGE: EFFICIENCY TRUMPS SPEED AND ALTITUDE (1976–1996)

Formation of three NASA and USAF digital fly-by-wire (FBW) testbed aircraft. The flexibility of digital FBW enables configuration changes for the testbeds more quickly and economically than analog FBW systems. From left to right are the McDonnell Douglas F-18 HARV (High Angle of Attack Research Vehicle); the Rockwell/MBB X-31 joint U.S./German project employing paddles in the exhaust stream of its single turbofan engine to vector engine thrust for increased agility; and the Lockheed Martin NF-16D Vista (Variable Inflight Stability Aircraft) currently used by the USAF Test Pilot School. Digital FBW is now fully accepted in operational use, a testimony to these one-of-a-kind testbed aircraft. (NASA)

Digital computers were initially developed during World War II to solve complex mathematical problems, and were used strategically to aid in decoding complicated encrypted enemy messages. Each World War II–era computer was generally purpose-designed with electronic circuitry for the specific problem it was designed to solve. These machines were large, sometimes filling a complete room, and required extensive cooling because they used vacuum tubes, which were state-of-the-art at the time.

From "Electronic Brain" to Super-Computer

Digital computers improved in the postwar period and their capabilities increased with greater speed for performing calculations as well as having more storage, allowing them to tackle even more complex problems. A major improvement was that computers could be designed to allow their programming to change to solve different problems without having to build a new computer.

These programming instructions spelled out the equations the computer was to solve, instructions on where data was stored to solve the equations, and instructions to the computer, which were written out on 80-column punch cards. These cards used what looked like a teletype machine, and the original memory devices were "ropes" of magnetic cores that were eventually replaced by magnetic tape drives and eventually disk drives.

Computers of the 1950s and 1960s had increased in capability by orders of magnitude beyond the World War II systems but were still large, expensive, and nonportable, and required extensive cooling. With the dawn of the space age in 1957 and the creation of NASA with its mission of sending a manned spacecraft to the moon by 1969, it was soon realized that manned spacecraft required some sort of onboard digital computer to assist the human crew for a trip to the moon. It also became obvious to the air force that the Minuteman ICBM upper stage warhead bus also needed an onboard digital computer to determine precisely when to release each of its 3 (and later 12) individual nuclear warheads.

In the 1950s, with development

of the transistor, aircraft avionics began to shrink in size. With the development of microcircuits using printed circuits on silicon, and which today we think of as chips, it became possible to create lightweight digital computers that did not require huge amounts of electrical power, and did not require as much cooling as earthbound computer rooms. This in turn led to digital computers that became increasingly wider deployed and more flexible, leading to the computer revolution we enjoy today.

The first aerospace vehicle to have an onboard digital computer was the Gemini spacecraft. The main purpose of this computer was to make calculations to allow the Gemini to rendezvous with another spacecraft, a capability that would be essential for the Apollo spacecraft going to the moon. The Gemini computer weighed 57 pounds, drew 95 watts of power, and had a whopping total of 12K of memory. The Apollo computer was similar in size, but had been upgraded to 2K of Random Access Memory (RAM) and 36K of the Read Only Memory (ROM). By today's standards these numbers may seem absolutely ridiculous, but these "stone age" computers successfully got men to the moon and back for the first time in 1969.

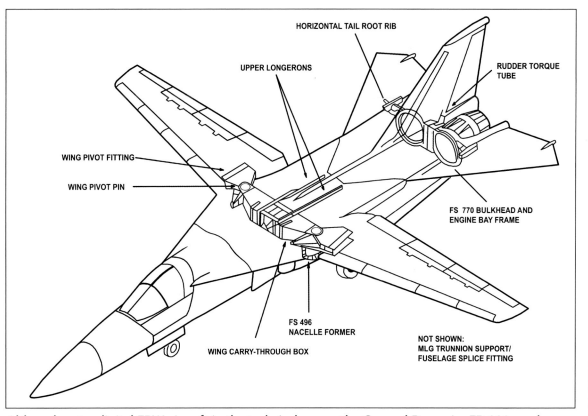

Although not a digital FBW aircraft in the technical sense, the General Dynamics FB-111 used a variable-geometry wing to achieve a wide range of flight performance parameters. This diagram depicts the F-111 with wings extended showing the two individual pivot points that allow a fully articulated wing-sweep motion. Wing pylons holding ordnance or external fuel tanks pivoted in conjunction with the aircraft's wing-sweep angle to achieve and maintain the absolute minimum amount of frontal drag. (Graphic digital enhancement by Craig Kaston)

The McDonnell Douglas F-15A air-superiority fighter first flew in 1972, and unlike the F-16, the F-15 is not a true FBW airplane with mechanical linkages to its control surfaces. F-15s have been in service for nearly 40 years and it is only in the past 5 years that the airplane, which is still being sold to foreign countries, received a FBW flight control system funded by the Royal Saudi Air Force, a current F-15 customer. This jet fighter has never been shot down by an enemy aircraft in aerial combat and maintains a combat ratio of 144:0. (USAF)

The U.S. Air Force investigated a digital FBW system on an experimental McDonnell Douglas F-4 Phantom testbed. The idea was to reduce the vulnerability of jet fighters to battle damage to its mechanical control linkages by replacing them with wires, which were lighter and considered less susceptible to destruction. The FBW Phantom flew shortly before NASA's Vought F-8 Crusader, but it was a close race. The Phantom was later modified with canard control surfaces mounted on the aircraft's air intakes to investigate unusual aerodynamic configurations, and presently resides at the National Museum of the U.S. Air Force in Dayton, Ohio. (USAF)

This F-16B wearing a classic flight test color scheme is in a perfect vertical climb high over Edwards AFB. The F-16 is designed to be highly agile and as a result, used what was known as a "relaxed stability" configuration where the center of gravity was farther aft than normal to allow quicker response to pilot commands for high-pitch rates. It was the first American fighter to use FBW, but the computers controlling it were analog computers rather than digital because it was designed in the early 1970s. (Chris Ledet via Mike Machat Collection)

Digital Fly-by-Wire Flight Controls: The Legacy of Apollo

With development of the jet engine and supersonic flight, high-performance aircraft required hydraulically actuated flight controls because their high speeds made pilot movement of the controls beyond human physical strength. The shift in aerodynamic characteristics in supersonic flight due to the shift in center of lift on a wing required stability augmentation systems (SAS) to help the pilot fly the airplane with what is known as "carefree handling." The U.S. Air Force, having learned hard lessons from aerial combat

Analog versus Digital Computers

Beginning in the 1930s, analog computers could be designed to monitor the analog values of pressure and velocity and take action to move the flight controls of an airplane to keep its behavior within specified limits. For each new airplane, a completely new analog computer might have to be designed. The automatic pilot invented in the 1920s used gyroscopes to monitor the attitude of an airplane and keep the airplane flying at a constant bank angle or heading if the pilot removed his hands from the controls or was otherwise distracted.

Through the 1960s, aircraft control functions used analog systems to assist the pilot in flying the airplane. Most aircraft were designed with some stability built in to ease the task of the human pilot when he took control. Designing an inherently unstable airplane often resulted in a high accident rate. Northrop's flying-wing designs of the 1940s had poor stability compared to conventional airplanes and required an analog autopilot with a yaw damper to help the human pilot. The advantage of an autopilot/yaw damper was that it never got tired or distracted, and once it was perfected, it was able to be used extensively on long-range flights.

In the digital world, real-world analog data is divided into samples, each one of which has a specific numerical value. In the early days of digital fly-by-wire, each of these samples was translated by a device known as an "analog-to-digital converter" into digital binary form, a mathematical system using ones and zeros to indicate a value. The digital computer took binary data and mathematically manipulated it to provide output, which was also a digital numeric value.

Human reaction time is approximately 240 ms (0.25 second) from hearing a command to taking action such as pressing a button or throwing a switch, assuming the human is expecting a command and knows what to do when he/she hears it. With modern digital computers, reactions can be considerably faster in correcting a situation. This means that any change in a condition will be detected within approximately 20 ms, and with proper programming the computer can respond much more quickly than a human. Most flight control software for digital fly-by-wire operates in a 50- or 60-hertz sampling rate, so therefore, it is possible to make an unstable airplane actually seem stable because it never seems to deviate from the condition the pilot is commanding, because the self-correcting digital system works so rapidly.

A further advantage of digital computers in aircraft is that all the old dials and gauges can be replaced by large, flat computer screens that can interpret digital data streams and produce a better display of graphic information to the pilot. This is known as the "glass cockpit," and today even 40- and 50-year-old airplanes can be retrofitted with computer screens to present digital data to the pilot. Airplanes built with all digital systems can have tailored screens that show data pilots were never able to see in earlier aircraft, easing pilots' workloads during complex high-stress situations such as approaches in bad weather or at night. Effort is ongoing by NASA and the manufacturers to ensure that data presented to the pilot is represented in such a way that "information overload" does not become a problem.

Digital computers rely heavily on their software to manipulate data to provide proper output, and even a small programming error can lead to disastrous results. With the increased complexity of tasks digital computers are performing, great care must be taken to ensure that programming instructions are still valid for every new task. A classic example occurred on the first flight of the European Ariane 5 space booster when an old program from the Ariane 4 was used to control the digital autopilot during launch. The Ariane 5 was more powerful than the Ariane 4 for which the program had originally been developed, and as the rocket accelerated it quickly exceeded the speeds for which the original program had been designed. The digital autopilot became confused because it no longer had a valid data stream, and the rocket began to veer off course. The range safety officer had no other choice than to destroy it.

The lesson that is continually relearned in aviation is that it's the small details that can kill you. The B-2 Spirit stealth bomber was a digital fly-by-wire airplane with 106 computers. In its first combat usage in 1999, a software flaw appeared that resulted in precision guided bombs occasionally hitting the ground miles away from where they were intended. An exhaustive investigation followed, which after two weeks discovered the problem. Two lines of software were added and two lines were relocated within the program. Software has become one of the major drivers of aircraft avionics costs due to the care that must be taken in programming digital computers and then validating and verifying the changes that have been made. Digital computers deliver great capabilities, but the cost of ensuring that the correct software is installed must be factored in.

The third generation of stealth technology (F-117 was first; B-2 second) is represented here by Lockheed Martin's F-22 Raptor air-superiority fighter. Capable of achieving supersonic flight without the use of afterburners, or "supercruise," the F-22 is a digital marvel in that it can control all the tactical activity within 150 miles of airspace while simultaneously communicating with other multi-mission fighters to completely dominate the aerial battlefield environment. (USAF)

Lockheed Martin, in partnership with CALSPAN Aeronautical Systems, created the one-of-a-kind NF-16D VISTA, or Variable Inflight Simulator Aircraft. Currently in operational usage by the USAF Test Pilot School, this advanced digital aircraft can be programmed while in flight to perform like almost any other aircraft in the Air Force inventory to give pilots and flight test engineers real-time airborne input on various aircraft they will be flying and testing in their careers. (Tony Landis)

This Lockheed X-35 Joint Strike Fighter Prototype Demonstrator is inflight over Edwards AFB during its flight test competition with the Boeing X-32 JSF candidate in 1999. After an intense and thorough flight test evaluation program, the Lockheed aircraft emerged victorious, and in a strange jump in designations ("F-24" was the next number in rotation) became the F-35 Lightning II. A fully digital aircraft from initial design through manufacturing and in flight, the F-35 has 9 million lines of software code compared to the 1980s-era Northrop B-2 that employs 1.6 million lines. (Mike Machat)

Approach and Landing Test (ALT) flights of the Space Shuttle Orbiter Enterprise at Edwards Air Force Base were required to prove the vehicle could be landed safely after re-entry before actual orbital space flights were conducted. The Enterprise was an atmospheric research vehicle only, designed to investigate the Orbiter's landing characteristics during final approach. It never actually flew in space. This is the moment of launch of the fourth free-flight test, the first time the Orbiter was ever flown without its protective streamlined tail cone. (NASA)

Four of the five ALT flights landed on the 13-mile-long Rogers Dry Lake. During Free Flight Five, its first attempt at the high-gain task of landing on the concrete runway at Edwards in front of a high-profile crowd, the vehicle experienced lateral oscillations that nearly drove it off the runway. This is known as pilot-induced oscillation (PIO) which had not occurred before when landing on the huge lakebed runways. (NASA)

experiences in Vietnam, decided that the era of the dogfight was not yet over, and the next generation of jet fighters should be highly maneuverable with their high-thrust jet engines being used to perform tighter turns and faster climbs, rather than simply accelerating to Mach 2 and beyond.

The McDonnell Douglas F-15 and General Dynamics F-16 were prime examples of this 1970s fourth-generation jet fighter. Both were optimized for air-to-air combat using missiles as well as an internal cannon, and both were expected to operate at high angles of attack, which could produce aerodynamic problems due to their swept wings. The small, lightweight F-16 was designed with a fly-by-wire flight control system, although computers in its initial flight control system were analog, and not digital.

Since the flight controls were receiving instructions via electrical wiring, it was possible to use a sidestick controller, which relied on pressure from the pilot's hand rather than stick position to communicate to the analog computers what performance was required. Although the control surfaces were hydraulically powered, they were controlled by electrical signals. There was no mechanical connection between the pilot's control stick and the control surfaces.

The Rockwell/MBB X-31 demonstrator was intended to evaluate using vectored engine thrust along with a digital FBW control system to allow controllable flight at extremely high angles of attack. Two X-31s were built and 550 test missions were flown over a period of 10 years. Test results showed that combining thrust vectoring with traditional flight controls would indeed allow the aircraft to remain under control when the wing was almost completely stalled, one test maneuver actually being carried out at a 70-degree angle of attack. Unfortunately one of the aircraft was lost in 1993 during a test mission over Edwards. The pilot ejected safely. (NASA)

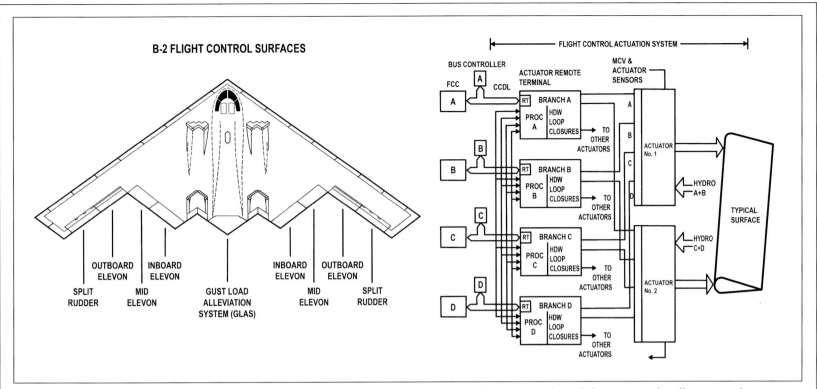

B-2 FLIGHT CONTROL SURFACES

SPLIT RUDDER — OUTBOARD ELEVON — MID ELEVON — INBOARD ELEVON — GUST LOAD ALLEVIATION SYSTEM (GLAS) — INBOARD ELEVON — MID ELEVON — OUTBOARD ELEVON — SPLIT RUDDER

FLIGHT CONTROL ACTUATION SYSTEM

BUS CONTROLLER — MCV & ACTUATOR SENSORS — ACTUATOR REMOTE TERMINAL

FCC — CCDL — BRANCH A / PROC A / HDW LOOP CLOSURES — TO OTHER ACTUATORS — ACTUATOR No. 1 — HYDRO A+B — TYPICAL SURFACE

BRANCH B / PROC B / HDW LOOP CLOSURES — TO OTHER ACTUATORS — HYDRO C+D — ACTUATOR No. 2

BRANCH C / PROC C / HDW LOOP CLOSURES — TO OTHER ACTUATORS

BRANCH D / PROC D / HDW LOOP CLOSURES — TO OTHER ACTUATORS

This plan view of the aircraft (left) shows its trailing-edge control surfaces. The schematic technical diagram (right) illustrates the Northrop B-2 Stealth Bomber's quadruple-redundant digital flight control system. (Graphic digital enhancement by Craig Kaston)

Grumman's X-29 was designed and built using composite wings with a 30-degree forward sweep, as well as a close-coupled canard surface for increased maneuverability. A digital FBW flight control system was installed because the aircraft's configuration combined with the location of its center of gravity well aft produced an aircraft so unstable (by 35 percent when the original intention had been only 20 percent), that without an FBW flight control system, the time to double amplitude of any disturbance was 8 milliseconds, rendering the airplane totally unflyable by a human pilot without the assistance of a digital computer.

The F-15 was a larger and heavier airplane with similar performance to the F-16, but the F-15 retained a mechanical connection between the pilot's normal center-mounted stick and control surfaces. Thus, it was possible to fly the airplane in the event of a complete power failure using the mechanical backup system, although without SAS, the flight envelope was somewhat limited. One change in the flight controls of both fighters was the use of independently controlled all-moving horizontal tail surfaces (called tailerons or rolling stabilators) to bank the airplane during high-angle-of-attack turns rather than using ailerons on the wing. This allowed the fighters to be more responsive to pilot control inputs, as well as an aerodynamically "clean" wing that produced more lift during high-angle-of-attack maneuvering.

An additional feature of the F-16 was that at high angles of attack, the forebody strakes on the forward part of the fuselage (looking somewhat like a hooded cobra snake, leading to the Fighting Falcon's unofficial nickname of Viper) generated lift at high angles of attack via vortices spilling off the forebody, allowing the aircraft to turn even more tightly. In order to increase agility even further, the F-16's center of gravity was moved farther aft than in earlier-generation fighters, with its digital flight control system keeping the aircraft from going unstable in pitch. This aft positioning of the center of gravity, essentially creating an unstable airplane, is known as "relaxed stability," and the advanced jet fighters of today employ it to enhance maneuverability.

Meanwhile, NASA was looking at using modern digital computers to assist in designing fighters with carefree handling. Several surplus Apollo spacecraft digital computers were acquired for the Dryden Flight Research Center and a surplus navy F-8 Crusader was chosen as the testbed to demonstrate the practicality of using a digital flight control computer to replace the analog computer in airplanes such as the F-16. The Apollo digital computer was limited by modern standards in terms of speed and memory capacity, but it soon became obvious that that was not a major problem because all airplanes basically followed the same laws of flight dynamics. Once the flight control program was written, it was seldom necessary to make major changes.

Because digital fly-by-wire architecture included no physical connection between the pilot's control stick and the flight controls themselves, everyone had a vested interest in ensuring that the software being used for a first flight was as error-free as possible. The air force was interested in fly-by-wire using analog computers, but initially there was hesitation on removing the manual flight control connection between the pilot's controls and the physical surfaces. The air force modified an F-4 Phantom to a fly-by-wire configuration with the rationale that using wires from the cockpit rather than rods and pulleys would be lighter and would improve the survivability of the Phantom in combat.

Digital fly-by-wire systems generally have the same controls as older jet aircraft. The difference is that when the pilot moves the controls, there is no physical connection to the control surfaces; he is merely signaling to a digital computer what he wants the airplane to do. The computer then sends signals to actuators for the flight controls,

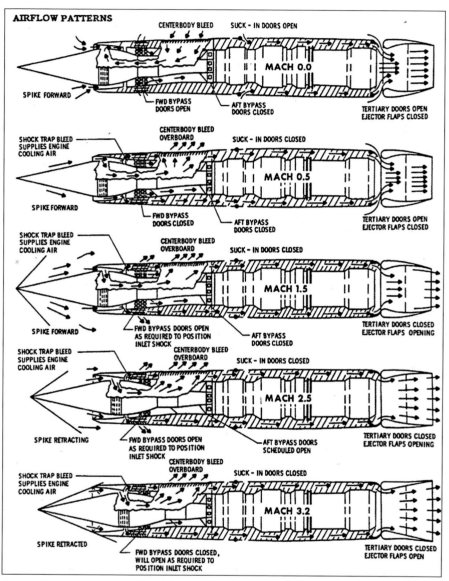

Reconfiguration of the SR-71 inlet as a function of Mach number. (The Mach number setting is printed on the cutaway of each nacelle.) The nacelle was designed for optimum performance at Mach 3.2 with the spike as far aft as it could go and all the doors closed for maximum pressure recovery and minimal external drag. The aircraft could go faster but now the inlet could not respond to the higher speed by moving the spike farther aft, leading to eventual inlet unstarts and possible flameout of the Pratt & Whitney J58 turbo-ramjet engine. (USAF)

The advent of airplanes employing stealth technology provided challenges to flight control systems because some of the configurations were somewhat unorthodox. One such aircraft was the Lockheed F-117 Nighthawk, which looked like a faceted mountain from the front but had a 60-degree wing sweep, not for speed, but to control radar reflection from the leading edge of the wing. It used an analog FBW system borrowed from the F-16, which apparently was satisfactory for an airplane that spent most of its life at low angles of attack. These photos were taken from a McDonnell Douglas KC-10A tanker aircraft. (Mike Machat)

causing the airplane to respond the way digital models in the computer think the airplane should respond.

Sensors monitor what the airplane does, comparing it to what the digital model predicts. If corrections are necessary, the computer automatically generates additional commands and then monitors the new response. An advantage of digital fly-by-wire is that instructions can be given to the computer not to allow flight beyond certain limits such as G-load, angle of attack, sideslip, speed, etc., so that the pilot does not have to be constantly looking at gauges in the cockpit to ensure he is not exceeding the aircraft's limits.

In the F-16, the pilot can pull hard on the sidestick controller without fear of inadvertently exceeding angle of attack or G limits. More modern digital fly-by-wire systems also control the engines. The Lockheed F-22's engines, as well as those of several new Russian jet fighters, also have variable vectored-thrust nozzles, which move to achieve tighter turns in a dogfight without the pilot having to worry about moving a separate nozzle lever, as was the case in the British 1960s-vintage Harrier.

Today's Lockheed F-35B Lightning 2 jet fighter goes even further than the F-22 by allowing the pilot to make vertical takeoffs and landings using just the control stick to choose a landing point. The aircraft's digital computer controls not only the attitude and speed of the fighter, but also automatically controls the thrust of the single jet engine and the attitude, considerably simplifying a vertical landing. This feature helps explain why the F-35 now has 9 million lines of software code to efficiently perform its mission versus the 1980's B-2 bomber design, which uses only 1.6 million lines.

All this capability comes at a price, however. Because there is no longer a physical connection between the pilot and the control surfaces, extreme reliability is required for the fly-by-wire system. Testbed aircraft could use individual computers as long as there was a manual backup control system for the pilot to fly the airplane should the computer fail. Operational aircraft, however, cannot afford a catastrophic failure, so at least one and sometimes as many as three backup systems were installed to not only provide redundancy, but also to compare results and vote out a failed system. This required all the computers to synchronize with one another to be using the same time standard for comparison.

Today, it is widely accepted that fly-by-wire aircraft technology is the wave of the future. The increased precision of measurement of digital quantities actually allows improved efficiency in operating complex systems. Early jet engines used mechanical fuel controls with cams, springs, and levers to control the engine because vacuum tubes could not withstand the temperatures around the jet engine. Eventually, with the development of solid-state technology, fuel controls became EECs (electronic engine control). Today, virtually all jet engines use DEECs (digital electronic engine controls). The flexibility inherent in computer-controlled engines allows increased efficiency, which, especially in the commercial engine field, is becoming a main influence on jet engine performance.

The digital fly-by-wire F-8 Crusader NASA tested in the 1970s demonstrated its value in an unexpected way. The United States developed a winged spacecraft as a follow-on to the Apollo program. A digital flight control system was chosen to allow the Space Shuttle Orbiter to land like a glider when returning from Earth orbit. A quadruple-redundant flight control system was to be used with a fifth computer monitoring the behavior of the other four systems. The orbiter *Enterprise* was built strictly to evaluate performance of the proposed space vehicle within the Earth's atmosphere, and was launched from the back of a Boeing 747 at approximately 20,000 feet to perform the final stage of the landing approach to the runways at Edwards AFB.

Performance seemed adequate when landing on the 10-mile-long lakebed runways but the fifth and final test was to land on Edwards' 15,000-foot concrete runway that modeled the runway being built at Cape Canaveral for operational shuttle landings. To make things more interesting, Prince Charles, heir to the Throne of England and himself a pilot, was present to witness this landing. A white line was painted on the runway indicating where the orbiter was to touch down. This is known as a "high gain" piloting task with the pilot wanting to touch down at exactly the right point. No problems had been experienced during the lakebed landings, but now pilot inputs resulted in erratic behavior in pitch and roll, which almost led to the orbiter going off the runway.

This behavior is known as "pilot induced oscillation" (PIO) and in analog systems was generally due to the design of the airplane itself, its stability, or lack thereof. As part of the investigation of the cause of the PIO, the fly-by-wire Crusader was reprogrammed to emulate the orbiter's fly-by-wire flight control system. Once the modification was made, the Crusader exhibited similar PIO tendencies when landing on the Edwards runway. Filters were then installed in the digital flight control system to smooth its behavior, leading to the elegant touchdowns seen in subsequent space shuttle landings.

In the 1980s, NASA and the U.S. Department of Defense were interested in exploring the advantages of a forward swept wing rather than the traditional aft swept wing for transonic flight. The Grumman X-29 was designed and built using composite wings swept forward at 30 degrees as well as a close-coupled canard surface for increased maneuverability. A digital fly-by-wire flight control system was installed because the configuration of the airplane, plus the location of its center of gravity well aft, produced a basic design that was so unstable that without a fly-by-wire flight control system the time to double amplitude of any disturbance was 8 ms, rendering it totally unflyable by a human pilot without computer assistance.

As a result, the X-29's digital flight control system was triple redundant and three analog back-up systems were installed. The two X-29 testbeds flew 294 successful test missions and both are now in museums. The experience gained in tailoring composite material fiber lay-up plus the maturity exhibited by the digital fly-by-wire system on such an unstable airplane greatly benefited the aviation industry and allowed a better analysis of forward-swept wings. The test results also showed that the theoretical advantages in maneuverability were vastly overstated, which explains why few airplanes today have forward-swept wings.

The Rockwell/MBB X-31 demonstrator was intended to evaluate using vectored engine thrust along with a digital fly-by-wire control system to exhibit controlled flight at extremely high angles of attack. Two X-31s were built, and over approximately 10 years a total of 550 test missions were flown. Test results showed that combining thrust vectoring with traditional flight controls indeed allowed the aircraft to remain under control when the wing was almost completely stalled, one test maneuver actually being carried out at a 70-degree angle of attack. Unfortunately, one of the aircraft was lost in 1993 during a mission over the dry lakebed at Edwards. The cause of the loss was a classic case of what is known in the computer world as GIGO or "garbage in, garbage out."

The air data system for the airplane had iced-up in unusually cold temperatures at altitude. Based on the incorrect air data output to the sensors, the flight control system essentially went berserk, refusing to respond correctly to pilot inputs. The pilot ejected as the errant aircraft went through high-G maneuvers that would have rendered him unconscious, and he would have died in the wreckage. The loss of a fly-by-wire B-2 stealth bomber on takeoff at Guam a decade later had a similar GIGO cause in that the air data system was blocked by moisture and the computers' calibration routines that are run on start-up took the incorrect data and assumed it was correct, thereby causing the airplane to leave the ground some 20 knots below actual takeoff speed.

The bomber's flight control system detected the spurious data during the takeoff roll and switched to alternate sensors, but too late, as the airplane was already airborne at too slow a speed. The bat-wing bomber sank back to earth once it flew out of the ground effect cushion at an altitude of 85 feet, both pilots fortunately surviving by using their ACES II zero-zero ejection seats. (Both the X-31 crash and B-2 Guam crash videos can be seen on YouTube.) The lesson is that digital fly-by-wire systems are a "two-edged sword." Tremendous benefits are gained along with hidden dangers.

World's Fastest Airplane Goes Digital: SR-71 Conversion in the 1980s

The Lockheed SR-71 reconnaissance aircraft was first designed in 1959 and originally used analog flight control computers to aid the pilot in flying at Mach 3. One of its biggest problems was controlling the air inlet flow to the Pratt & Whitney J58 turbojet engines at cruise speed. Although analog computers were used in the initial inlet design, it was one the most troublesome aspects of operating the SR-71. The inlet had to reconfigure itself at various speeds to control the complex system of shockwaves in the inlet that allowed Mach 3+ air to slow to Mach 0.5 prior to entering the engine. This complex

inlet is what actually produced over half the net propulsive thrust of the Blackbird at Mach 3 and 85,000 feet. The inlet functioned almost like a ramjet in front of the turbojet engine with the pressure buildup in essence pulling the airplane while the jet engines' thrust pushed it.

The problem was that the inlet was very sensitive to angle-of-attack changes, which altered the complex shockwave pattern to the point that the final normal shock was expelled from the inlet. Called an "unstart," this resulted in the loss of much of the net thrust in that nacelle within a second. The engines and their inlets were mounted far outboard on the wing due to bow-shockwave geometry at Mach 3, and with the sudden loss of thrust that far out on the wing, the airplane yawed violently, disturbing the shockwaves on the other wing and resulting in the inlet on that side also unstarting. An analog automatic system helped the pilot in trying to recapture the shockwaves after the unstarts but it was often too slow to prevent violent yaw and deceleration.

It was assumed that the SR-71 would be replaced by something flying faster and higher in the late 1980s, but that was not to be. An SR-71 was loaned to NASA as a research testbed for high-speed flight and a digital flight control system was retrofitted using a single set of computers in the late 1970s. Initial results from the relatively low-priority flight test program indicated there could be advantages in using digital computers in the operational SR-71 fleet to solve a number of problems, mainly the engine inlet unstarts that plagued the Blackbird fleet since its beginnings in the early 1960s. Twelve SR-71s had been lost in crashes during its operational and flight test career, and there were a number of pilot issues involved in controlling an airplane at very high speeds and very high altitudes.

A contract was awarded to Lockheed and Honeywell to install a triple-redundant digital flight control system in the SR-71 fleet that not only commanded the flight controls but also controlled the complex integrated inlet and propulsion system that allowed the aircraft to cruise at Mach 3 and beyond. An SR-71 had actually been torn apart in flight in the mid-1960s when an unstart occurred during a 30-degree bank turn, with both crewmembers departing the aircraft due to the violence of the yaw and subsequent breakup while their ejection seats actually remained in the cockpit. The pilot survived his long fall from 80,000 feet thanks to his full pressure suit and automatic parachute opening at 10,000 feet, but the back-seater unfortunately died of a broken neck suffered during the breakup.

Several other SR-71s were lost due to pitch-ups at low speeds and heavy weights while coming off the tanker with its long forward fuselage and chine acting like a canard wing. This exceeded the control authority of the elevons to arrest the pitch-up. It was thought that a digital computer could both aid the flying characteristics of an airplane as well as more precisely control the complex engine inlet. The flight test aircraft substituted digital pressure transducers mated to short air data lines for the dozens of feet of pneumatic piping inherent in the analog 1959 design, and three modern (for 1979) solid-state digital computers replaced the analog systems.

Northrop's B-2 flying wing bomber was another stealth design, based on a previous Northrop flying wing aircraft from the 1940s with a very checkered reputation. The B-2 employed a quadruple-redundant digital flight control system and underwent thousands of hours of flight simulator study prior to its actual first flight. It exhibits "carefree handling" in the sense that the computers did not allow the pilot to command the airplane into uncharted aerodynamic regions and firmly refused to go there, despite pilot inputs from the cockpit. (Mike Machat)

The system became known as DAFICS (Digital Automatic Flight and Inlet Control System), but one difference from modern digital systems was that the original mechanical connections to the flight controls were retained because it was easier to leave them in than to remove them. So in the event of a triple computer failure, the pilot could still fly the airplane manually, as had been possible for more than 20 years, but without the benefit of a stability augmentation system (SAS). The SR-71 was extremely difficult to fly at high Mach, especially in turns, and in the frequent training simulator practice sessions it was not unusual for the pilot to lose control of the aircraft and crash with a failed SAS.

A year was spent using the dedicated flight test airplane (S/N 61-7955) to debug the system software and adjust settings so that the DAFICS airplanes flew almost identically to the original analog airplanes. During these tests the biggest initial challenge was reducing inlet unstarts. The flight test team tracked the number of unstarts in the initial developmental test and counted more than 50 unstarts during the debugging process, fortunately none as violent as the one that resulted in the destruction of the earlier Blackbird. The original quest to reach Mach 3 in 1962 had also been somewhat of an empirical process, and it took more than six months of incremental steps and modifications of the inlets and J58 engines to proceed from Mach 2.6 to the Mach 3.2 cruise speed.

Once DAFICS was certified it was soon discovered that the increased precision of measuring parameters using digital systems allowed the inlets to respond to disturbances so quickly that inlet unstarts became a thing of the past. To familiarize new crewmembers with the characteristics of an unstart, the instructor had to take manual control of the inlet and deliberately mishandle the configuration to induce it. DAFICS had finally cured the problem that had been the biggest concern for the SR-71 since its inception.

Test pilots for North American's triplesonic XB-70 Valkyrie experimental bomber mentioned that paying passengers on a future supersonic airliner would never fly on it again if they experienced an inlet unstart; now it seemed that the digital era had improved inlet operation so that this would probably be a non-issue. The SR-71 fleet was withdrawn from service by the air force within five years of being modernized with digital equipment in both DAFICS and avionics before further refinements could be evaluated. Limited flight test use seems to indicate that with the increased digital precision, it would probably have been possible to expand the SR-71's range at Mach 3 by making software adjustments to the inlet schedule that reduced specific fuel consumption of the J58 engine.

The only supersonic trainer left in the U.S. military inventory today is Northrop's T-38A Talon, a Mach 1.4 turbojet-powered aircraft that first flew in 1959. Still an attractive aircraft, the jet is used today for pilot training, safety chase, and pilot proficiency missions in support of the Northrop B-2 and Lockheed U-2. Note the forward-angled air intakes as originally designed. (Tony Landis)

Close-up of the new-style subsonic air intake on a Northrop Grumman T-38C, a modified version of the Talon that features this high-efficiency intake and a digital "glass cockpit" that is more compatible with aircraft in the current air force fleet. The T-38's classic "white rocket" color scheme has now also been replaced with a two-tone gray-green medium-visibility paint job. Protective anti-FOD covers are shown installed over the intakes. (USAF/AFFTC)

Fly-by-Wire: All Things Are Now Possible

Before the digital age, engineers went to great lengths to design an airplane's shape and weight distribution as well as a control system the pilot could use to control that aircraft. Today, new airplanes can be produced with shapes that defy any previous rules of aeronautical design. The basics of lift, drag, and thrust must still be respected in designing an airplane. But now liberties can be taken with the balancing of those forces to produce a flyable aircraft despite its shape. Good examples of this are two radical stealth airplanes: the Lockheed F-117 and Northrop B-2. Both designs would not have been possible without the development of modern avionics and digital flight control computers. The F-117 features a highly swept wing not for high Mach flight, but to position the radar signature of the airplane in such a way that enemy radars will not see it.

The faceting of the fuselage has the same function, but the combination of the two features results in aerodynamics that make the airplane unflyable without computer assistance. Before the age of digital assist, the single pilot spent all his time making sure the airplane did not crash during his night mission rather than searching for the target on his sensor screens. Similarly, the B-2 is a large flying-wing aircraft with unusual flying characteristics that pilots sometimes struggled with, and which negated the effectiveness of a large airplane, which was all wing. The hundred-plus computers on the modern B-2 allow a two-person crew to do what a five- or six-person crew used to do, while at the same time operating an aircraft that some pundits predicted could never fly. Now there's talk of building even larger flying wings to transport cargo and perhaps passengers in an environmentally friendly manner in the future.

Commercial Aviation in the Digital Age: Upgraded Avionics and Smaller Flight Crews

During the Cold War, the military became involved in advanced digital electronics out of necessity. Fly-by-wire allowed advances in high-speed jet fighter operations for aerial combat. By comparison, commercial jet operations involved transporting large numbers of people and cargo from point A to point B at high subsonic speeds while showing a profit. As a result, jet airliners did not initially require digital flight controls, but the development of wide-body passenger jets plus the new economic viability of overnight delivery with jet freighters placed a huge demand on commercial aviation that was not present in the early 1950s.

Boeing and Airbus, the two surviving mega-corporations building jetliners, were constantly in competition with one another for sales. This fact, coupled with deregulation of the American airline industry in 1978, produced many advances in jet airliners as well. Success in commercial aviation was measured by filling seats in airliners at the lowest practical price while meeting increasing demand for global passenger travel. Modern digital technology allowed digital fly-by-wire cockpits manned by only two pilots who were seated facing multiple display screens rather than the banks of dials and gauges that formerly required a crew of from three to five, thus saving quantum amounts of money for the airlines.

Larger, 100,000-pound-thrust high-bypass turbofans dwarfed the Boeing 747's original 50,000-pound-thrust JT9 or CF6 engines of the 1970s and enabled twin-engine wide-bodies to operate long distances over water. Airbus was the first airliner manufacturer to use digital fly-by-wire technology in their twin-engine A320, but Boeing eventually conceded the advantages of the new technology and adapted digital "glass cockpits" for their 757 and 767 twin-jets. Even the 747 went digital with development of the advanced 747-400.

A more representative airliner of the 21st Century was the Boeing 777 series. First flown in 1995 this was Boeing's first digital fly-by-wire product and also had two jet engines instead of four, but could carry almost as many passengers as the early 747s over longer distances and at lower fuel costs. The 777 never had an accident until 2013 when an accident at San Francisco International Airport (SFO) resulted in three deaths on a bright sunny day when the airplane hit a seawall on landing. All three pilots in the cockpit were studying the new digital avionics on the airplane rather than visually flying the jet, leading to stories questioning whether the new Boeing's cockpit was too complicated. Nevertheless, the 777 is actually Boeing's most profitable airplane, and more than 1,400 have been delivered out of a total order book of nearly 1,900.

The 777 incident at SFO illustrates a problem that modern aircraft with carefree handling and digital avionics have experienced. Airline pilots' basic flying skills may become rather rusty when digital avionics do most of the flying via the autopilot. An airline pilot in the 1990s was quoted as saying that airline management did not want the pilots hand-flying the aircraft, because the autopilot's handling of flight controls and engines produced better fuel mileage than the human pilot, leading to lower fuel costs. However, when unexpected situations occur and the autopilot disengages for safety reasons, the pilot may not be ready to analyze the situation and take corrective action fast enough to avert disaster.

The loss of an Air France Airbus A330 over the South Atlantic due to unusual thunderstorm activity and air data sensors icing-up (sound familiar?) was traced to the fact that the two pilots on the flight deck were pulling on their sidestick controllers in opposite directions trying to correct the perceived problem. This accident happened at night and in bad weather, and perhaps might not have occurred if the pilots had been able to see the horizon. However, this loss of pure piloting skills has been a factor in a number of recent aviation accidents, and in a related issue, may be relevant on a much grander scale if there's ever a mass deployment of self-driving cars on the highways.

Cockpit state-of-the-art, circa 1948. The Lockheed 749 Constellation sports all the latest equipment and instrumentation. The flight engineer sits sideways behind the captain and co-pilot, and is responsible for the aircraft's systems and powerplant operation. Note the P-38–style control yokes and small pans atop the glareshield containing alcohol, which was ignited inflight to provide heat and prevent icing on the windshield. (Wings & Airpower Historical Archive)

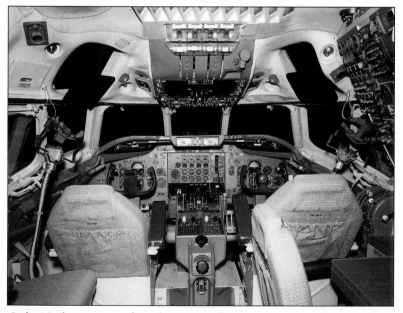

Only 10 short years after the Constellation came America's first-generation of jetliners and this Douglas DC-8's cockpit. Gone are such piston-era devices as "prop synchronizers" and "engine analyzers." New features such as "eyebrow windows" above the windshield enhanced cockpit visibility. Throttles became "thrust levers" and the flight engineer was now responsible for cabin and cockpit pressurization at altitudes of up to 39,000 feet. (Douglas Aircraft Company)

First-generation "glass cockpits" made their appearance with the Boeing 757, 767, and Airbus A320 jetliners in the 1980s. The next step was the Boeing 747-400 and McDonnell Douglas MD-11 (shown). Although Airbus pilots used sidestick controllers, as did pilots of the F-16 jet fighter, Boeing and McDonnell Douglas stayed with traditional control yokes to maintain commonality with their other company products. Flat digital display screens replaced round "steam" gauges as well.

In this beautiful and graphic depiction of wing vortices and low-pressure condensation, an Air New Zealand Boeing 777-300ER (for Extended Range) gently descends the ILS for Runway 24 Right at LAX on an unusually stormy day in Los Angeles. Note the conical vortices emanating from the outboard corners of the flaps, and not the wingtips. (Jean-Louis Delezenne)

Chapter 8

WHERE HAVE ALL THE RECORDS GONE?

Aviation artist Mike Machat was assigned to document the takeoff of the transcontinental SR-71 record flight and created this dramatic painting for the U.S. Air Force Documentary Art Program. In the darkness of the predawn takeoff, he initially thought the SR-71 had three landing lights until he realized that, as the aircraft drew closer, the two "outboard landing lights" were actually monstrous afterburner plumes seen through the fronts of the engines. The author and his daughter were also present for this takeoff at Palmdale, California, that morning. (Mike Machat)

What if airplanes set world records and nobody cared? What if those records no longer captured major headlines in the media or caused any mention in the aerospace industry's leading trade journals? As unbelievable as these questions may seem, that is almost where we are today. Emphasizing this compelling point even further, as these words are being written the first round-the-world flight by a solar-powered airplane was completed. Little mention was ever made on the nightly news, however, and most ironically, this flight took longer to complete than the Douglas World Cruisers' similar epic journey in 1924.

The Last Hurrahs: Records as the Cold War Ends

By the mid-1980s, digital technology was beginning to reshape society in the form of desktop computers, which soon began to appear in even the most modest homes around the world. This in turn led to wholesale usage of the Internet, which tied computers together and allowed digital applications that once would have been science fiction. Relatively few speed or altitude records were set that decade, with the exception of a publicized attempt to fly around the world nonstop without refueling.

The *Voyager* took off in December 1986 from Edwards Air Force Base where supersonic flight first occurred, and landed at the same location nine days later having flown around the world at the equator at an average speed of 116 mph using two piston engines powering propellers, although for most of the flight one of the engines was shut down to conserve fuel. More than 3,500 reporters were at Edwards for the takeoff and watched as the overloaded airplane dragged and then scraped off its wingtips on the 15,000-foot runway, fortunately without major damage to the rest of the aircraft.

A crowd of 55,000 spectators, plus innumerable world media representatives, were present for the landing beamed around the world

The Burt Rutan–designed Voyager *aircraft is skirting ominous storm clouds during its historic globe-circling nonstop unrefueled flight of December 1986. This unorthodox twin-boom, twin-engine, two-place aircraft had an impressive lift-to-drag ratio of 37:1, comparable to a high-performance competition sailplane, and nearly twice that of the average jet airliner today. After this record-breaking flight, the aircraft was donated to the Smithsonian National Air & Space Museum where it is displayed for visitors today. (Artwork copyright Craig Kodera)*

Lockheed SR-71 Aircraft Commander Lt. Col. Ed Yeilding under-goes final operational checks on his David Clark S1030 full pressure suit in the Life Support Shop at Air Force Plant 42, Palmdale, California. The crew was in final preparation for their epic record-breaking cross-country flight on 6 March 1990. (Joyce Baker via Tony Landis Collection)

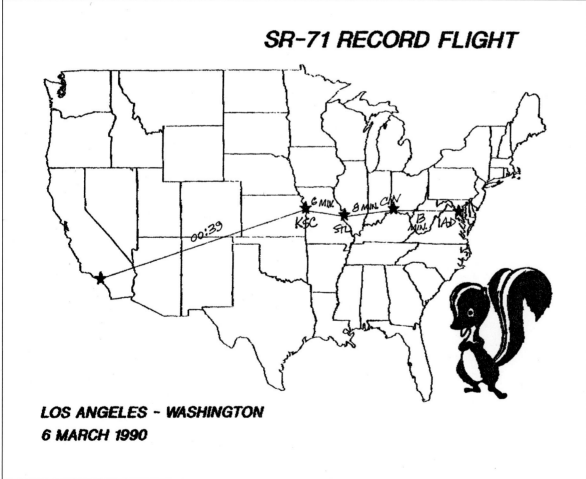

SR-71 RECORD FLIGHT

LOS ANGELES ~ WASHINGTON
6 MARCH 1990

Showing both the planned route of flight and the times required to transit each leg of that route, this promotional map was distributed to members of the media who observed the historic takeoff from Palmdale, California, beginning the record flight. Audible gasps could be heard from media members during the pre-flight briefing as Lt. Col. Ed Yeilding announced the seemingly miniscule amounts of time it would take to fly from one checkpoint city to the next, with perhaps the most dramatic pairing being only 39 minutes between Los Angeles and Kansas City.

The waiting crowd at Washington's Dulles International Airport was given a special treat with the SR-71's high-speed low pass over Runway 01 Right in full afterburner, after crossing the United States from coast to coast in just 68 minutes. The average speed between the start and finish gates was Mach 3.2 (2,100 mph), or 1 mile every 1.2 seconds. The original flight plan called for this pass to be performed at 1,500 feet, but low ceilings across the Eastern Seaboard prevented that from happening. (Tony Landis)

live via satellite. The two-person crew included Jeanna Yeager (no relation to Chuck) as well as Dick Rutan, older brother of Burt Rutan, the engineer who designed the airplane. *Voyager* was made of composite materials to save weight, but was very lightly stressed. Since it was unpressurized, it had to fly at low altitude to avoid flying near thunderstorms that could cause structural failure due to turbulence.

The flight controls were manual, but despite the presence of an autopilot, the handling characteristics of the airplane were not very good. Nevertheless, the "ultimate record," as it was called, had finally been accomplished: flying around the world in a single flight unrefu-

eled. That flight's celebrated success also capped off a year that had started tragically with the loss of Space Shuttle *Challenger* in January.

Voyager's record was broken in 2005 by a single-pilot jet version of the aircraft called *Global Voyager*, which had a single turbojet engine mounted above the mid-fuselage in a similar fashion to the Heinkel 162 *Salamander* of World War II. Like its propeller-driven brethren, *Voyager*, it had an exceptionally high lift-to-drag ratio of 37:1, akin to a racing sailplane, and was also lightly built. Since it had a jet engine, it was able to generally fly above the weather and avoid turbulence.

The flight took 2 days and 19 hours, taking off and landing at the Municipal Airport in Salina Kansas. *Global Voyager* set two other long-distance unrefueled records before being retired to the National Air and Space Museum in Washington, DC, being displayed not far from its propeller-driven brother as well as Wiley Post's Lockheed Vega, *The Winnie Mae*, in which he made the first solo around-the-world flight in 1933.

Another amazing national speed record was set in March 1990 when a Lockheed SR-71 Blackbird made its final flight to retirement at the Smithsonian National Air and Space Museum, setting a new transcontinental speed record of only 68 minutes (from a start gate in California to a finish gate in Maryland). The aircraft's average speed was more than 2,100 mph, or 35 miles per minute! The flight's progress across the United States was reported on morning television

Shown taking off at dusk from Runway 25 Right at Los Angeles International Airport, this American Airlines Boeing 777-300ER serves as the backbone of today's long-range global airline fleet. The 777 was the world's first airliner designed and built using digital technology (the B-2 was the world's first aircraft created that way), and entered airline service in 1996. Much to Boeing's surprise, the success of this twin-engine aeronautical marvel has actually hastened the end of production of the four-engine 747 series. (Jean-Louis Delezenne)

Here's an example of how contemporary manufacturing techniques can affect an aircraft's design: Before Boeing's fluid-looking blended wing body (BWB) seen on the next page, came the "Spanloader," a large-scale McDonnell Douglas concept in 1979 where the aircraft's wing carries the payload and the fuselage carries the fuel. With the majority of aircraft structure at right angles to each other and with minimal aerodynamic fairing, you can clearly see the advantages of computer-aided design combined with composite structure and fully digital manufacturing techniques. (Mike Machat)

NASA's Boeing X-48 unmanned aerial vehicle (UAV) represents the blended-wing-body configuration for a possible full-scale aircraft of the future. Based on a design proposal from Great Britain's Cranwell Aeronautical University, Boeing has proposed variations of the X-48 as a 200-foot-span flying wing to carry passengers and cargo. The success of Northrop Grumman's B-2 flying wing with its digital flight controls indicates that a similar control system could be installed for a full-scale version of the X-48 as a viable design for future commercial aircraft. (NASA)

network news, but only several hundred spectators were on hand for the predawn takeoff from Air Force Plant 42 at Palmdale, California.

FAI observers were stationed at FAA Air Traffic Control Centers along the route of flight to make it an official record, and phone calls were received at these facilities as the sonic boom was heard on the ground. When the callers were notified that it was a Mach 3 sonic boom from the Blackbird making its final flight, rather than complaints, the calls turned into requests for more information. Unlike earlier record-setting flights that were greeted with bands, generals, and military decorations, the crowd to greet the Blackbird's arrival at Dulles International Airport was made up mainly of media people with very few air force officers in attendance. The air force crew was comprised of Lt. Col. Ed Yielding and Lt. Col. J. T. Vida, both of

whom had flown the airplane operationally before joining the Palmdale flight test detachment. Vida was the all-time SR-71 high-time flyer with more than 1,600 hours in the jet.

The aircraft flown to the museum was fully representative of an operational SR-71, although cameras were replaced by souvenirs being flown at the request of numerous people who supported the airplane during its 25-year career. Yielding reported that the flight was fairly routine, although problems with the fuel system after its first aerial refueling initially caused some concern. Yielding emphasized that any qualified SR-71 crew could have set the record, although to set the official records, Yielding had to obtain an FAI sporting license, which he dutifully carried in the pocket of his pressure suit during the flight. Unknown to most people was that one of the SR-71's biggest supporters was Sen. John Glenn, who had set the first supersonic coast-to-coast record in 1958 before he became an astronaut. He enthusiastically supported the SR-71's record run, considering it a fitting salute to the fastest jet aircraft ever built.

NASA continued to fly a handful of SR-71s on research missions, but 9 October 1999, the last Blackbird made its final flight at the Edwards Air Force Base air show, delivering a final Mach-3 sonic boom on all 250,000 spectators. The era of faster and higher that began with the invention of the jet engine was now history. It was only fitting that the final flight demonstration be made in front of a huge throng of spectators at an air show in Southern California, almost 90 years after the Dominguez Hills air show first helped trigger the American public's interest in aviation.

Aviation Records in the 21st Century

Although aviation records today seldom make the newspapers, they are still being made. On the FAI website are detailed instructions on how you can set an aviation record yourself. Most of the records are under FAI category C, which is heavier-than-air aircraft, although not exclusively; category A for balloon flight occasionally also records new records. Most of the records are for point-to-point flights in less elapsed time than any other pilot has achieved. These are sometimes known as "vanity records," but occasionally some occur that are noteworthy of aviation progress.

On 10 November 2005, a Boeing 777-200LR (Long-Range) flew from Hong Kong to London piloted by a Boeing commercial test pilot for a straight-line distance record for jet aircraft in class C1 T of 11,663 nautical miles (21,600 km) unrefueled. The flight took 22

Perhaps the finest example of blending mature and advanced technology is Boeing's magnificent 747-8, the latest (and final) version of the legendary 450-passenger jumbo jet. With the first 747 having flown in 1969 and more than 1,500 built to date, this airline industry veteran set the standard for mass air travel from the 1970s through the 1990s. The 747-8 version features a two-pilot digital cockpit, 7,700-mile range, and uprated General Electric GEnx-2B67 turbofans capable of producing 66,500 pounds of thrust each. (Jean-Louis Delezenne)

Civilian Record Setters in the 20th Century

In this closing chapter, mention must be made of special record holders from the past who did not have Machmeters on their instrument panels, but who advanced aviation by setting records in the grand tradition of Sir Edmund Hillary who, when asked why he climbed Mt. Everest, replied succinctly, "Because it was there." These bold civilian airmen and women set numerous records in single-engine and twin-engine light aircraft, helicopters, and even sailplanes, all in the quest of showing the world that it is the human spirit coupled with innovative technology that triumphs in aviation.

The pilots who set these records in the 1950s and 1960s were unique individuals, coming from all walks of life, backgrounds, and experience levels. People like the first modern women pilots; grandfathers flying single-engine airplanes halfway around the world; jet helicopter, and even glider pilots all got in on the act, setting impressive records in exclusive categories with many different types of aircraft. Pictured here are just four of these record-setting aircraft and mention of the pilots who flew them.

Only three months after its maiden flight, the turbine-powered Sud-Est SE 3130 Alouette II established a world altitude record for helicopters of 26,932 feet on 6 June 1955. Two years later, this revolutionary French helicopter broke its own record by climbing to 36,027 feet. Both flights were piloted by Sud Chief Test Pilot Jean Boulet, proving the value of turbine powerplants for helicopters. The Alouette II was the world's first production jet helicopter. (Republic Aviation Corporation)

A rugged and reliable Aero Commander 560 (such as the one pictured here) was flown by renowned aviatrix Jerrie Cobb on her many world-record flights for speed and distance, plus an altitude mark of 37,010 feet. A flight instructor at age 19, Cobb advanced to become the first woman to fly in the Paris Air Show, and the first woman to qualify in meeting the requirements for a NASA astronaut during the Mercury Program, although she never flew in space. She was a pilot and manager for Aero Commander, becoming one of the first women executives in aviation.

The sleek Piper PA-24 Comanche was flown by Max "The Flying Grandfather" Conrad to establish a new world distance record for light single-engine aircraft of 7,668 miles flying non-stop from Casablanca, Morocco, to Los Angeles, California. The flight was made from 2 June to 4 June 1959. On 24 November 1959, Conrad flew a 180-hp version of the Comanche as seen here from Casablanca to El Paso, Texas, in an elapsed flight time of 56 hours, a class record that still stands to this day.

This customized high-performance Schweizer SGS 1-23E flown by NASA Flight Test Center Director Paul Bikle set the world altitude record for sailplanes on 25 February 1961 climbing to 46,629 feet in the powerful Sierra Wave over Bishop, California. Although the ultimate altitude record for sailplanes became 49,009 feet in 1986 and 50,720 feet in 2006, Bikle's record for maximum altitude gain by a sailplane (an impressive 42,303 feet) has never been broken.

hours and 22 minutes and set two new speed records for that size aircraft for L.A. to New York of 610 mph and New York to London of 566 mph.

This vanity record was set to commemorate the delivery of a brand-new Boeing twin-jet airliner, demonstrating that it could fly more than halfway around the world carrying passengers. A team of nine pilots were on board but the pilot in command was Boeing test pilot Captain Suzanna Darcy Hannemann. She is a chief test pilot for Boeing and the first woman test pilot to qualify in the Boeing 747-400. The record was recognized not only by the FAI, but also the Guinness book of records, which had declared 9 November as "set a record day" throughout the world.

Perhaps it was just a coincidence, but on the same date in 1935 a female aviator named Jean Batten took off to fly solo from England to Brazil in a single-engine Percival Gull, for which she later received the FAI gold medal for the year. Her airplane still exists and hangs in the main terminal building of the Jean Batten International Terminal at the Auckland International Airport in New Zealand, her home country. In contrast, the record-setting 777 was subsequently delivered to Pakistan International Airlines (PIA) and is still flying today.

Absolute records in category C are much more difficult to set because it is difficult to exceed the previous records by the required margins. Nevertheless, aviation technology is progressing to the point where as this book was being written, a solar-powered airplane called *Solar Impulse 2*, flown by a single pilot, was making its way at all of 40 mph on the first round-the-world flight by a solar-powered airplane (category CS in the FAI taxonomy).

The flight originated in Abu Dhabi in March 2015, but a hiatus occurred when the electric motors overheated near Japan, requiring six months to repair the damage. *Solar Impulse 2* only generates 30 to 40 hp, very similar to that of the cumbersome gasoline engines that set the first FAI records in 1909. Its wingspan, however, matches the 747 to accommodate all the solar cells necessary to provide that 30 to 40 hp. It is to be hoped that *Solar Impulse 2* records can be broken as decisively as Wiley Post accomplished less than a decade after the first globe-circling flight in 1924 (see the Preface for more information.)

In the 21st Century emphasis has shifted to cleaner and "greener" aviation to minimize the impact on the environment. The aviation industry and notably NASA are concentrating on methods of increasing fuel economy, reducing noise produced by airplanes, and perhaps achieving supersonic flight without sonic booms. The aircraft engine industry has produced geared turbofans, promising fuel savings of 4 to 5 percent over existing engines by allowing the fan to rotate independently of the compressors at an optimum speed for maximum thrust. This does not sound like much, but considering there are some 100,000 commercial airline flights per day throughout the globe, single-digit savings could produce carbon emission reductions of thousands of tons per year, not to mention savings on the amount of fuel consumed in the millions of dollars.

The turbofan initially reduced jet engine noise, and geared turbofans may produce even quieter engines to the point where attention is now being paid to reducing aerodynamic airflow noise in addition to engine noise. Efforts to mask aircraft noise from people on the ground has led to construction of the X-48 blended wing body UAV for exploration of aerodynamic efficiency as well as sound suppression prior to building full-scale vehicles. Unusual configurations of vehicles are now being proposed because fly-by-wire digital flight control systems are an accepted means of making such aircraft practical. NASA is conducting research that potentially could be as significant as the streamlining investigations of NACA back in the 1930s.

Who Needs Pilots Actually Onboard the Aircraft?

An FAI category that increasingly will be populated by records is category "U" for unmanned aerial vehicles (UAVs), or drones as the press seems to use more frequently. Records have already been set with large UAVs, such as the Global Hawk flying from the United States to Australia and back nonstop, but the explosion in popularity of personal drones virtually guarantees that more UAV records will be established, and probably in increasingly bizarre categories. In 2016 a mockup of a single-person UAV was exhibited at the Consumer Electronics Show in Las Vegas. The vehicle was intended to be an on-call (à la Uber) personal transportation vehicle that does not have a human pilot but is controlled by the passenger's iPhone to deliver him or her from the pickup location to the destination.

The competition for longest distance flown without a pilot, but with a human aboard, could set off the ultimate "vanity records" race from 42nd and Broadway to the Staten Island Ferry Terminal in New York. Just as the television program "Robot Wars" was popular for some time, perhaps "Robot Air Races" might catch on, complete with virtual reality transmissions to the spectators' iPhones. This could in turn lead to improvements in microchips and batteries, as well as more radical aerodynamic configurations. (The quad rotor configuration of many personal drones is very similar to the original design of the first helicopters, which failed miserably at the beginning of the 20th Century.) New configurations could lead to faster times on the race course and new speed records in the U category.

History has a way of repeating itself, often in ways the originators of a technology never dreamed of. When John Moore-Brabazon flew that pig in 1909, little did he know that in less than 100 years, half the population of the earth would be flying routinely at 600 mph, and on giant airplanes without propellers on more than 100,000 flights a day. Aviation technology led to the space-age, which led to the miniaturized computer age, which in turn led to the UAV age, not to mention the Internet. Who knows what the next step will be? But whatever it is, there will probably be more aviation records to be set and more races and cash prizes to be won. And watch out for that flying car. It may not have a human driver.

Northrop Grumman's Global Hawk Unmanned Aerial System (UAS) is fast becoming the mainstay of the intelligence, surveillance, and reconnaissance (ISR) community worldwide. Technically considered a non-combat aircraft, this autonomously piloted vehicle has been used for both military and civilian humanitarian purposes, and is capable of remaining airborne at subsonic speed for more than 30 hours at a time. The Global Hawk truly represents the ultimate application of digital technology in today's aerospace industry. (USAF)

GLOSSARY

A

Aerodynamic: Properties referring to an aircraft in flight

Afterburner: Injects raw fuel into engine exhaust for greater thrust

Area Rule: Reduces aircraft structure surface area to facilitate supersonic flight

B

Ballistic: Unpowered flight resulting from initial high-velocity thrust

Barograph: Old-style air pressure-driven flight data recording instrument

BLCS - Boundary Layer Control System

Boundary Layer: Thin sandwich of air that adheres to an aircraft in flight

C

Chord: Distance between a wing's leading and trailing edges

Coffin Corner: Area of flight envelope with highest risk to pilot

Compressor Stall –Explosion resulting from air-starvation of a jet engine

Control Authority: Ability of a control surface to maintain positive control

Critical Mach Number: Speed at which airflow over a wing reaches Mach 1

D

Decompression: Loss of cockpit or cabin pressure at high altitude

Density Altitude: Relative air density reflecting ambient altitude and temperature

DFC: Digital Flight Control System

Dive Brakes (see Speed Brakes)

Dorsal Fin: Vertically oriented control surface above the fuselage

Drag Chute: Large parachute that deploys to slow an aircraft on landing

Drogue Chute: Small parachute that deploys to stabilize an aircraft or body

E

Ejection Seat: Device that ejects pilot from an aircraft in an emergency

Escape Capsule: Device that fully encloses pilot for ejection from aircraft

Escape Pod: Portion of aircraft that ejects the crew during an emergency

Exo-atmospheric: Flight outside the Earth's atmosphere

Explosive Decompression: (see Decompression) Violent decompression

F

FAA: Federal Aviation Agency

FAI: Federation Aeronautique Internationale

Fairing: Aerodynamic surface to streamline airflow and reduce drag

FAR: Federal Air Regulation

Flame-out: Condition of jet engine losing thrust due to air or fuel starvation

Fowler Flap: High-lift device that extends to enlarge wing area for slower flight

Full Pressure Suit: Pneumatic suit worn by pilots and astronauts to protect them from depressurization above 50,000 feet

G

Glass Cockpit: Refers to digital instrumentation data presented on flat screens

GPS: Satellite-based Global Positioning System for pin-point navigation

H

Head-Up Display: Projection of flight data onto screen in line of sight

Hypersonic: A speed faster than Mach 5, or approximately 3,500 mph

Hypoxia: Condition brought about by lack of oxygen at high altitudes

L

Lakebed: Large, flat desert surface for high-performance flight OPS

Laminar Flow: Optimal airfoil shape for high-performance flight

Leading Edge: The front edge of a wing or control surface

Leading Edge Flap: High-lift control surface mounted on front of wing

Leading Edge Slat: High-lift surface mounted on front of wing

M

Mach: The speed of sound, named for Austrian physicist Ernst Mach

Mach Tuck: Tendency of a supersonic aircraft to depart controlled flight

Mock-Up: Non-flight-capable full-size model of an aircraft or structure

N

NAA: National Aeronautic Association

Nacelle: Streamlined housing for an aircraft engine

O

Official Record: An aviation record sanctioned by the FAI

OPS: Operations

Ordnance: Weaponry carried under or aboard an aircraft

P

Partial Pressure Suit: Suit with mechanical devices worn by jet pilots to protect them from depressurization above 50,000 feet

Pitot Tube: Small cylindrical device that measures an aircraft's speed

Pylon: Structure connecting an engine nacelle to the wing

R

RCS: Reaction Control System

Rocket: A liquid or solid-fuel motor that does not need oxygen for thrust

Rocket Sled: Track-mounted, rocket-propelled, high-speed ground test vehicle

S

SAS: Stability Augmentation System

Shockwave: Fast-moving air mass that surrounds a supersonic aircraft

Sonic Boom: Thunderous noise when a shockwave strikes the ground

Space: An altitude higher than 50 miles above the Earth's surface

Span: Lateral distance from the wing root to the wingtip

Span-wise Flow: Boundary layer airflow that moves toward the wingtip

Speed Brakes: Drag-producing devices that deploy into the airflow

Strake: Horizontal blade to guide airflow over control surfaces

Subsonic: Slower than Mach 1, or the speed of sound (650-720 mph)

Supersonic: Faster than Mach 1, or the speed of sound (650-720 mph)

T

Theodolite: Precision radar tracking device for recording flight test data

Trailing Edge: The aft edge of a wing or control surface

Transonic: Speeds at or near the speed of sound (Mach 0.8-1.2)

Turbojet: Engine that produces thrust from compressing an air/fuel mixture

Turbofan: Turbojet engine augmented by larger fan blades either fore or aft

U

Unofficial Record: Set by an aircraft that did not meet rules for an official record

V

Ventral Fin: Vertically oriented control surface below the fuselage

Vortex Generator: Small pod or blade-like device that activates airflow

W

Wing Fence: A vertical blade mounted on a wing to control span-wise flow

Wing Root: Portion of wing that mates with an aircraft's fuselage

Winglet: Small wingtip-mounted vertical fin that reduces drag

Y

YAPS Boom: Records aircraft's Yaw, Angle-of-attack, Pitch, and Side-slip

Yaw Vane: A weather vane type of device mounted on a YAPS Boom

APPENDIX ONE: SPEED RECORDS

Progressive Speed Records In Aviation History

These may not have been sanctioned by FAI.

1. First Speed Record Faster than 100 mph: 100.23 mph, Deperdussin Monoplane, Jules Vedrines, Pau, France, 22 February 1912
2. First Speed Record Faster than 200 mph: 210.64 mph, Nieuport-Delage 29, Sadi Le Cointe, Villesauvage, France, 25 December 1921
3. First Speed Record Faster than 300 mph: 313.59 mph, Macchi M-52 Seaplane Racer, Maj. Mario de Bernardi, Venice, Italy, 6 November 1927
4. First Speed Record Faster than 400 mph: 407.02 mph, Flt. Lt. G. Stainforth, Supermarine S.6B, Ryde, England, 29 September 1931
5. First Speed Record Faster than 500 mph: (Estimated to be German Messerschmitt Me-262 at 541 mph; there are no official records for this speed being attained for the first time during World War II.)
6. First Speed Record Faster than 600 mph: 603.0 mph, Gloster Meteor IV, Sqdn. Ldr. P. Stanburg, Moreton-Valence, England, 19 October 1945
7. First Speed Record Faster than 700 mph: 715.15 mph, North American F-86D, Lt. Col. William Barnes, Salton Sea, California, 16 July 1953
8 First Speed Record Faster than 800 mph: 822.27 mph, North American F-100C, Col. Horace Haynes, Edwards AFB, California, 20 August 1955
9. First Speed Record Faster than 900 and 1,000 mph: 1,132.0 mph, Fairey Delta 2, Peter Twiss, Chichester, England, 10 March 1956
10. First Speed Record Faster than 2,000 mph (Rocket): 2,175 mph, Bell X-2, Capt. Milburn G. Apt, Edwards AFB, California, 27 September 1956
10. First Speed Record Faster than 2,000 mph (Jet): 2,070.1 mph, Lockheed YF-12A, USAF Col. Robert Stephens, Edwards AFB, California, 1 May 1965
10. First Speed Record Faster than 3,000 mph: 3,074 mph, North American X-15, USAF Maj. Robert M. White, Edwards AFB, California, 21 April 1961
10. First Speed Record Faster than 4,000 mph: 4,093 mph, North American X-15, USAF Maj. Robert M. White, Edwards AFB, California, 9 November 1961

Major World Speed Records

2. Low-Level Closed-Course Speed: 753.4 mph, Douglas F4D-1 Skyray, U. S. Navy LCDR James Verdun, Salton Sea, California, 3 October 1953
3. Low-Level Absolute Speed: 755.1 mph, North American YF-100A Super Sabre, USAF Major Frank K. "Pete" Everest, Edwards AFB, California, 29 October 1953
4. Unofficial Speed: 670 mph (Mach 1.01), Bell X-1, USAF Capt. Charles E. "Chuck" Yeager, Edwards AFB, California, 14 October 1947
4. Unofficial Speed: 1,325.6 mph (Mach 2.005), Douglas D-558-II, A. Scott Crossfield (NACA), Edwards AFB, California, 20 November 1953
5. Unofficial Speed: 1,660.2 mph (Mach 2.44), Bell X-1A, USAF Capt. Charles E. "Chuck" Yeager, Edwards AFB, California, 12 December 1953
7. U.S. Transcontinental Speed: 6 hrs. 8 min. (592 mph), Boeing 367-80, Alvin M. "Tex" Johnston, Seattle (Boeing), Washington, to Washington DC, 16 October 1955
9. Unofficial Speed: 2,175 mph (Mach 3.2), USAF Bell X-2, Capt. Milburn G. Apt, Edwards AFB, California, 27 September 1956
10. High-Altitude Closed-Course Speed: 1,132 mph, Fairey Delta 2, Peter Twiss (Fairey), Farnborough, England, 10 March 1956
11. U.S. Transcontinental Speed: 3 hours and 23 minutes (Mach 1.1), Vought F8U-1P, Maj. John H. Glenn, NAS Los Alamitos, California, to NAS Floyd Bennett Field, New York, 16 July 1957
14. Woman's Absolute Speed: 1,312.4 mph (Mach 2), Dassault Mirage III-R, Jacqueline Auriol, Istres, France, 26 August 1959
15. High-Altitude Closed-Course Speed: 1,216.5 mph, Republic F-105B Thunderchief, USAF BGen. Joseph H. Moore, Edwards AFB, California, 11 December 1959
16. High-Altitude Absolute Speed: 1,525.9 mph (Mach 2.3), Convair F-106A, USAF Gen. Joe Rogers, Edwards AFB, California, 15 December 1959
20. Absolute Speed: 2,070.1 mph (Mach 3.0), Lockheed YF-12A, USAF Col. Robert Stephens and USAF Lt. Col. Dan Andre, Edwards AFB, California, 1 May 1965
21. Transatlantic Speed: 2 hours and 53 minutes (1,350 mph), BAC Concorde, Captain Leslie Scott, London to New York, 7 February 1996
22. Round the World Speed: 23 hours and 13 minutes (1,350 mph), Aerospatiale Concorde, Captain Claude Delorme, Lisbon, Portugal to Lisbon, Portugal, 13 October 1992
23. Round the World Non-Stop Speed: 36 hours and 13 minutes (631.16 mph), Rockwell International B-1B, During Operation "Coronet Bat," Lt. Col. Doug Raaberg, Dyess AFB, TX to Dyess AFB, Texas, 2 June 1995

Significant Speed Flights by Mach Number

Mach 1 (670 mph): Bell X-1 rocket-powered research aircraft in a dive, 14 October 1947, USAF Capt. Charles E. "Chuck" Yeager. The first flight to Mach 1.

Mach 1 (670 mph): Republic XF-91 turbojet-powered with rocket boost in level flight, 9 December 1952, RAC Test Pilot Carl Bellinger. The first flight to Mach 1 in level flight.

Mach 1 (670 mph): North American F-100A Super Sabre turbojet-powered aircraft, 25 May 1953, NAA Test Pilot George "Wheaties" Welch. The first flight to Mach 1 by an operational military aircraft.

Mach 2 (1,250 mph): Douglas D-558-II rocket-powered research aircraft, 20 November 1953, NACA Test Pilot A. Scott Crossfield. The first flight to Mach 2.

Mach 2 (1,250 mph): North American X-10 turbojet-powered unmanned research aircraft, 1955. The first flight to Mach 2 by a jet aircraft.

Mach 2 (1,250 mph): Aerospatiale Concorde, 4 November 1970, Test Pilot Andre Turcat. The first flight to Mach 2 by a commercial airliner.

Mach 2.41 (1,360 mph): Convair F-106A turbojet-powered interceptor aircraft, 20 November 1953, USAF Col. Joe Rogers. The fastest speed ever attained by a single-engine aircraft.

Mach 3 (2,094 mph): Bell X-2 rocket-powered research aircraft, 27 September 1956, USAF Capt. Milburn G. "Mel" Apt. The first flight to Mach 3.

Mach 3.2 (2,200 mph): Lockheed A-12 Blackbird turbo-ramjet-powered reconnaissance aircraft, July 1963, Lockheed test pilot Lou Shalk. The first flight to Mach 3 by a manned jet aircraft.

Mach 3 (2,020 mph): North American XB-70 turbojet-powered aircraft, 8 April 1966, Lt. Col. Fitzhugh L. "Fitz" Fulton, Jr. Largest aircraft ever to fly at Mach 3.

Mach 4 (2,275 mph): North American X-15 rocket-powered research aircraft S/N 56-6671, 7 March 1961, USAF Maj. Robert M. White. The first flight to Mach 4.

Mach 5 (3,603 mph): North American X-15 rocket-powered research aircraft S/N 56-6671, 23 June 1961, USAF Maj. Robert M. White. The first flight to Mach 5.

Mach 6 (4,093 mph): North American X-15 rocket-powered research aircraft S/N 56-6671, 9 November 1961, USAF Maj. Robert M. White. The first flight to Mach 6.

Mach 6.7 (4,520 mph): North American X-15A-2 S/N 56-6671, 3 October 1967, USAF Maj. William J. "Pete" Knight. The fastest flight of a manned, winged aircraft within the Earth's atmosphere (the space shuttle is considered a spacecraft rather than an airplane for the purpose of this book).

APPENDIX TWO: ALTITUDE RECORDS

Progressive Altitude Record Flights Above the following Altitudes

These may not be sanctioned by FAI.

10,000: 12,828 feet, Bleriot, Roland Garros, France, September 4, 1911

20,000: 20,060 feet, Nieuport 17, Gerard Legagneaux, France, 28 December 1913

30,000: 31,420 feet, Curtiss K-12, R. Rohlfs, New York, New York, 18 September 1919

40,000: 41,796 feet, Junkers, W. Neuenhofen, Germany, 25 May 1929

50,000: 51,364 feet, Caproni 113, Mario Pezzi, Italy, 8 May 1937

60,000: 60,000 feet, Bell X-1, USAF Capt. Charles E. "Chuck" Yeager, Muroc Flight Test Center (Edwards AFB), California, 26 May 1948

70,000: 71,902 feet, Bell X-1, USAF Maj. Frank K. "Pete" Everest, Muroc Flight Test Center (Edwards AFB), California, 8 August 1949

80,000: 83,235 feet, Douglas D-558-II, USMC Maj. Marion Carl, Edwards AFB, California, 21 August 1953

90,000: 90,440 feet, Bell X-1A, USAF Maj. Arthur "Kit" Murray, Edwards AFB, California, 26 August 1954

100,000: 126,200 feet, Bell X-2, USAF Capt. Iven C. Kincheloe, Edwards AFB, California, 7 September 1956

200,000: 217,000 feet, North American X-15, Edwards AFB, California, USAF Maj. Robert M. White, 11 October 1961

300,000: 314,750 feet, North American X-15, Edwards AFB, California, USAF Maj. Robert M. White, 17 July 1962

354,200: North American X-15, NASA Test Pilot Joe Walker, Edwards AFB, California, 22 August 1962 (Highest flight of a manned aircraft until the Space Shuttle in 1981.)

First Astronaut Wings Awarded to Aircraft Pilots

United States Astronaut wings were awarded to eight X-15 pilots who flew into "space," attaining an altitude of more than 50 miles (or

264,000 feet), as defined by the U.S. Air Force, and 62 miles (or 327,360 feet), as defined by NASA. These pilots are:

1. Maj. Robert M. White (USAF)
2. Joe Walker (NASA)
3. Maj. Robert A. Rushworth (USAF)
4. Jack McKay (NASA)
5. Capt. Joe Engle (USAF)
6. Bill Dana (NASA)
7. Maj. William J. "Pete" Knight (USAF)
8. Maj. Mike Adams (USAF)

Significant Altitude Record Flights In Aviation History

Reflecting the absolute frenzy of record-breaking flights from 1954 to 1963. Note the close proximity of many of the dates.

Helicopter Altitude: Sud-Est SE 3130 Alouette II turbine-powered helicopter sets altitude record of 26,932 feet in France, 12 March 1955

Absolute Altitude: British English Electric Canberra sets world altitude record of 65,876 feet for non-rocket-powered aircraft, 29 August 1955

Helicopter Altitude: J. E. Woman, reaches 30,335 feet in a Cessna YH-41, setting a world helicopter altitude record, 28 December 1958

Absolute Altitude: Grumman F11F-1F Super Tiger, piloted by LCDR George C. Watkins, sets world altitude record of 76,932 feet, 18 April 1958

Absolute Altitude: Rene Carpentier flies the French hybrid rocket and turbojet-powered Nord Trident interceptor to 78,452 feet, 2 May 1958

Absolute Altitude: Lockheed F-104A reclaims the altitude record when Major Howard C. Johnson pilots it to 91,243 feet, 7 May 1958
Time-to-Climb: Sea level to 40,000 feet, Douglas F4D-1, USMC Maj. Edward N. LeFaivre, USMC, China Lake, California, 22 May 1958

Helicopter Altitude: Sud-Est SE 3130 Alouette II turbine-powered helicopter sets altitude record of 36,027 feet in France, 13 June 1958

Absolute Altitude: 94,659 feet, Sukhoi T-431, Maj. Victor Ilyushin, Soviet Union, 14 July 1959

Absolute Altitude: McDonnell F-4 Phantom II, piloted by Commander L. E. Flint, sets an altitude record of 98,556 feet, 6 December 1959

Payload-to-Altitude: North American YA3J-1 Vigilante sets a world altitude record of 91,450.8 feet with a 1,000-kilogram payload, 13 December 1960

Absolute Altitude: USAF Capt. Joe B. Jordan, in a Lockheed F-104C Starfighter, sets an altitude record of 103,389 feet, 14 December 1959

Absolute Altitude: 118,898 feet, Mikoyan Ye-66A (MiG-25), Georgi Mosolov, Soviet Union, 28 April 1961

Sailplane Altitude: 46,260 feet, Schweizer 1-23E, Paul Bikle, 24 January 1961

Helicopter Altitude: Kaman H-43B Huskie helicopter sets altitude record of 32,840 feet for its class, 18 October 1961

Payload-to-Altitude: Major David W. Craw sets a world altitude record in a Boeing C-135B. He climbs to 47,171 feet with a 66,138-pound payload, 17 April 1962

Payload-to-Altitude: Grumman Albatross UF-2G sets a world altitude record for amphibians with a 1,000-kilogram load at 29,640 feet, 12 September 1962

Payload-to-Altitude: 5,000 kg to 85,364 feet, Convair B-58 Hustler, USAF Lt. Col. Fitzhugh L. "Fitz" Fulton, 14 September 1962

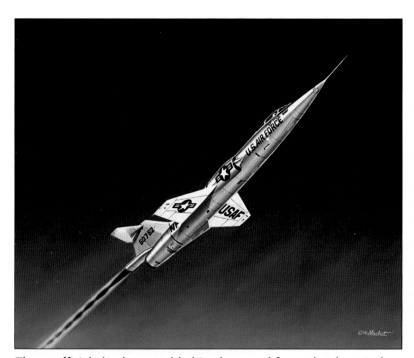

The unofficial absolute world altitude record for rocket-boosted turbojet aircraft was set by a Lockheed NF-104, flown by USAF Maj. Robert W. Smith to 120,800 feet over Edwards AFB, California, on 6 December 1963. (Mike Machat)

INDEX

N

NACA, 21, 56, 57, 62, 64, 66, 67, 69, 71, 72, 73, 74, 77, 78, 115, 116, 119, 126, 146, 182

NASA, 21, 55, 58, 74, 78115, 118, 120, 121, 127, 133, 134, 140, 141, 144, 146, 148, 149, 153, 154, 156, 157, 158, 160, 164, 166, 167, 178, 179, 181, 182

Nash, Capt. Jason Wade, 89

National Advisory Committee for Aeronautics, 21

National Aero Club trophy, 116

National Air Races, 88, 104

National Museum of the Air Force, 127, 146

NATO, 106, 114, 154

O

Operation Sun Run, 102, 104

Ortega prize, 31

P

Paris Air Show, 33, 143, 149, 154, 155, 181

Payne, Maj. William, 143

Peterson, CDR Forrest, 121

Popson, Capt. Ray, 74

Post, Wiley, 17, 24, 26, 27, 28, 31, 176, 182

Powers, Francis Gary, 144

Prince Charles, 166

Project Bullet, 101, 103

Pulitzer trophy race, 18, 19, 23, 68, 102, 104

R

Rahn, Bob, 83

Reims International Air Exhibition, 10, 12, 18

Rickenbacker, Capt. Eddie, 7

Ridley, Capt. Jack, 55, 62

Rogers, Cal Galbraith, 15

Rogers, Joe, 140

Rushworth, Col. Robert, 121

Rutan, Burt, 173, 175

S

Salamander, 44, 175

Salmon, Herman, 109

Sayer, Jerry, 39

Schneider Cup, 18, 20, 21, 23, 24

Schroeder, CPT Rudolph, 30

Scientific American trophy, 11

Selfridge, LT Thomas, 10

Setter, Retired Air Force Col. Lewis, 27

Skunk Works, 47, 48, 52, 55, 144

Smith, R. G., 65, 147, 150

Smithsonian National Air and Space Museum, 48, 173, 176

Solar Impulse 2, 182

Sperry, Lawrence, 88

Sputnik, 105, 106, 107

Stanley, Bob, 41, 43, 122

T

Taft, President Howard, 12, 15

The Question Mark, 98, 99, 100

The Right Stuff, 55

Thompson, Milt, 134

Tissandier, Gaston, 25, 30

Tissandier, Paul, 107

U

Udet, General, 37, 38

USS *Franklin D. Roosevelt* (CVB-42), 93

V

Verdin, LCDR James B., 83

Vida, Lt. Col. J. T., 178

Vin Fiz, 15

Voisin, Gabriel and Henri, 10

von Braun, Werner, 36

von Ohain, Dr. Hans, 33, 35, 36, 37

Voyager, 173, 175, 176

W

Walker, Joe, 72, 121

Welch, Anne, 30, 187

Wendel, Fritz, 38

Whitcomb, Richard, 109, 114, 115, 116, 142, 155

White, Al, 135

White, Major Bob, 120

Whittle, Frank, 33, 35, 39

Wilson, Capt. H. J., 55

Winnie Mae, 24, 26, 27, 28, 31, 176

Wright, Katherine, 11

Wright, Orville, 8, 9, 10

Wright, Wilbur, 9, 10

Y

Yeager, Cap. Charles E., 55, 58, 61, 62, 68, 71, 77, 79, 81, 117, 175

Yielding, Lt. Col. Ed, 178

Artist's depiction of the ascent phase of Maj. William J. "Pete" Knight's epic world-record flight in the North American X-15A-2 on 3 October 1967. Knight reached Mach 6.7 (4,250 mph), the fastest speed ever flown in the X-15, and the fastest achieved by a human pilot flying a winged aircraft within the Earth's atmosphere. This wordy description addresses the technicality of the Space Shuttle Orbiter having flown Mach 25 on orbit, and at approximately Mach 4 during re-entry while passing through 102,100 feet, the peak altitude of Knight's flight in the X-15A-2. (Mike Machat)

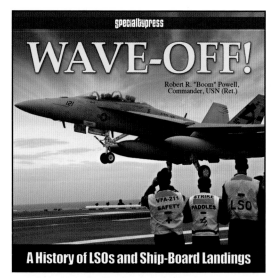

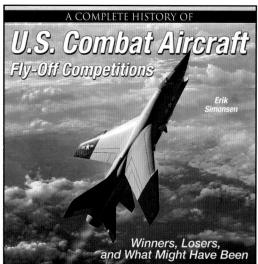

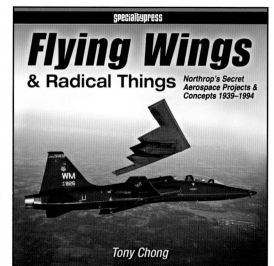

WAVE-OFF! A History of LSOs and Ship-Board Landings *by Tony Chong* This book tells the story of Landing Signal Officers from the first carrier operations in 1922 through World War II, the early jet era, Korea, Vietnam, and up to today's nuclear-powered leviathans. Also explained are naval aircraft and equipment development through the years; it covers both the faster and heavier aircraft and the changes in shipboard flight-deck systems. Diagrams showing the evolution of aircraft carrier deck design from World War I to the present are also included. Hardbound, 10 x 10, 192 pages, 188 b/w and 82 color photos. **Item # SP235**

A COMPLETE HISTORY OF U.S. COMBAT AIRCRAFT FLY-OFF COMPETITIONS *Erik Simonsen* Many advanced and now legendary aircraft have been designed, built, and flown in every generation of aviation development. Focusing on the Cold War era, this book shows readers how crucial fly-off competitions have been to the development of America's military air arsenal. This book explains in detail how fly-off competitions are conducted, and it shows what both competing aircraft designs looked like during their trials, and what the losing aircraft would have looked like in operational markings had it actually won. Hardbound, 10 x 10, 228 pages, 395 color and 156 b/w photos. **Item # SP227**

FLYING WINGS & RADICAL THINGS: Northrop's Secret Aerospace Projects & Concepts 1939–1994 *Tony Chong* John K. "Jack" Northrop and the company he founded in 1939, Northrop Aircraft, Inc., will be forever linked with the giant futuristic Flying Wings of the 1940s. Here for the first time is the untold story of Northrop's rare, unique, and formerly super-secret aircraft and spacecraft of the future. Featuring stunning original factory artwork, technical drawings, and never-before-seen photographs, this book shows an amazing array of radical high-performance aircraft concepts from Jack Northrop and his team of brilliant and innovative engineers. Much of this material has only recently been declassified. Hardbound, 10 x 10, 192 pages, 361 b/w and 70 color photos. **Item # SP229**

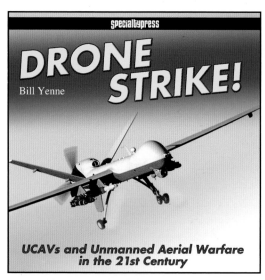

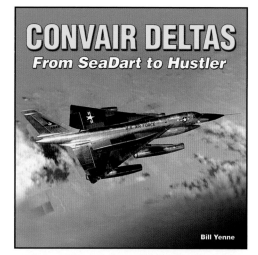

DRONE STRIKE!: UCAVs and Aerial Warfare in the 21st Century *Bill Yenne* Takes you from the end of the 20th Century through today's latest technical wonders, covering such amazing unmanned aircraft capabilities as aerial refueling and landing aboard aircraft carriers even more accurately than manned aircraft. Hardbound, 10 x 10 inches, 160 pages, 300 b/w and color photos. **Item # SP238S**

CONVAIR DELTAS: From SeaDart to Hustler *Bill Yenne* Now in paperback, this book tells the compelling story of America's aerospace industry in its heyday, when manufacturers boldly took the initiative to explore futuristic new designs by actually building and test-flying airplanes to determine how well, if at all, they would work. Softbound, 9 x 9, 216 pages, 344 b/w and color photos. **Item # SP231**

WORLD'S FASTEST SINGLE-ENGINE JET AIRCRAFT: Convair's F-106 Delta Dart Interceptor *Col. Doug Barbier, USAF (Ret.)* This book provides an insightful and in-depth look at the sixth member of the Air Force "Century Series" family of supersonic fighters. From initial concept through early flight test and development and into operational service, every facet of the F-106's career is examined. Hardbound, 10 x 10 inches, 240 pages, 350 photos. **Item # SP237**

Specialty Press, 838 Lake Street South, Forest Lake, MN. Phone 800-895-4585 & 651-277-1400 Fax: 651-277-1203
www.specialtypress.com
Crécy Publishing Ltd., 1a Ringway Trading Estate, Shadowmoss Road, Manchester, M22 5LH, England. Phone: 44 161 499 0024 Fax: 44 161 499 0298
www.crecy.co.uk